caring for
Cut Flowers

Rod Jones

Second Edition

LAND
LINKS

National Library of Australia Cataloguing-in-Publication entry

Jones, Rod, 1959– .
Caring for cut flowers.
2nd ed.

ISBN 0 643 06631 4

1. Cut flowers – Postharvest technology.
2. Cut flowers – Storage. 3. Cut flowers – Preservation.
4. Flowers – Postharvest technology. I. Title.

635.91568

Published by
LANDLINKS PRESS
150 Oxford Street (PO Box 1139)
Collingwood VIC 3066
Australia

Telephone:	+61 3 9662 7500
Fax:	+61 3 9662 7555
Email:	sales@publish.csiro.au
Web site:	www.publish.csiro.au

Set in 11/14.75 pt Meridien
Printed in Australia by Ligare

CONTENTS

ACKNOWLEDGEMENTS

The author gratefully acknowledges permission to use information provided by:

Professor Michael Reid
Department of Environmental Horticulture
University of California, Davis, USA

Dr John Faragher
Department of Natural Resources and Environment
Institute for Horticultural Development
Knoxfield, Victoria

INTRODUCTION

The aim of this book is to show florists how to make cut flowers last as long as possible. Longevity of flower vase life is the best value a florist can give to a customer. Flowers that wilt quickly do not provide customer satisfaction and only serve to deter people from buying flowers again. Market research has shown that if a bunch of flowers lasts, on the other hand, customers will be happy to buy flowers regularly. And most importantly, happy customers will return to the florist who supplied them with the best quality, longest lasting flowers.

Cut flowers are among the most perishable products you can buy. A branch of science known as flower postharvest physiology is dedicated solely to finding the most effective ways of maintaining flower quality and making flowers last. While correct flower care techniques will not magically transform poor quality flowers into first class material, a few simple, inexpensive procedures will make sure top quality flowers do not die within a few days. Poor flower care is an expensive business. You may pay top dollar for the best Lilies or Roses, but they will be dead within five days if you don't look after them properly.

Florist customers are also looking for information on how to maximise flower longevity. You can use the information contained in this book to advise your customers on what to do. Most people don't have a clue how to look after flowers, so anything you can tell them is worthwhile. The more they know, the longer their flowers should last and the more likely it is that they will buy flowers again, hopefully from you.

This book gives a brief introduction to the four basics of flower care: temperature, water, flower food and how to avoid problems with ethylene gas, which speeds up wilting and flower drop in certain flowers. It provides advice on buying and preparing for sale, tips on customer service, and techniques for floral arrangements. It then lists, in alphabetical order, specific instructions for around 150 flowers, including what to look for when buying, what temperature the flower is happy at, how to prepare flowers before arranging, whether they need flower food, whether they are sensitive to ethylene gas, and any other extra tips I've learnt over the years.

This Second Edition of *Caring for Cut Flowers* covers twice as many flowers as the First Edition and provides much more up-to-date information. Much of the new information and wider coverage is the result of contact with growers and vendors during my four years' experience as Quality Manager of the National Flower Centre, which is part of the Melbourne Markets, located at Footscray in Melbourne.

Chapter 1

THE FOUR BASICS OF FLOWER CARE

There are four basic factors that affect the vase life of cut flowers:

1 the temperature at which the flowers are kept;
2 the quality of the water in which they are placed, and their ability to take up that water;
3 the food they are given; and
4 their sensitivity to ethylene.

A little scientific information is necessary to explain how each of these factors affects flowers.

1. TEMPERATURE

Temperature is arguably the most critical factor in the care of cut flowers. Cut flowers are living organisms, and have only a limited life. Nothing 'goes off' more quickly than a cut flower on a hot day. Most flowers are made up of thin, delicate petals that lose water and wilt rapidly; and the warmer the weather, the faster flowers wilt.

Flowers are genetically 'programmed' not to last. It takes a great deal of energy for a plant to produce large, brilliantly coloured flowers, and the plant would 'prefer' to do this only for a short time. For this reason, flowers are mostly short lived; they are designed to be pollinated by bees, insects or birds, set fruit or seeds, and then die.

Humans have changed this scenario to a certain extent by breeding plants that will produce large colourful flowers continually and for as long as possible. An interesting example of this is the Pansy, which in the wild is a perennial (flowers every year), but these days lasts only one season as it tends to flower itself to death.

The energy needed for plants to produce flowers is supplied by photosynthesis using sunlight, carbon dioxide and water to produce sugars (see below), and from the water and minerals the plant gets from the soil. The process known as *respiration* converts these foods (mainly sugars) into energy.

Respiration is used by all living organisms to produce energy through the chemical breakdown of sugars, carbohydrates and fats. In flowering plants, the energy produced by respiration is used to produce flowers in the first place, then open buds, create colours, nectar and pollen, and finally, wilting and flower death. The by-products of respiration are water, carbon dioxide and *heat*. Flowering plants have a relatively high rate of respiration, which results in the flowers themselves generating a substantial amount of heat. In other words, flowers naturally generate their own heat. And, in most cases, this heat will cause cut flowers to die faster if not controlled by refrigeration.

Respiration itself is controlled by temperature. Higher temperatures result in a high respiration rate, which in turn speeds up flower development, which means the flower will die more quickly. A helpful way to understand this is to think of the respiration rate as the rate of deterioration, or ageing of the flower. Therefore, the higher the temperature the flower is held at, the faster the respiration rate, and the sooner the flower will die. This is why we need to keep most flowers cool. Research has clearly shown that carnations held at 10°C, for example, will age and die eight times faster than a flower held at 1°C.

Cool temperatures, on the other hand, drastically slow down flower ageing by slowing respiration. Cool temperatures have the added benefits of dramatically reducing water loss and the effects of ethylene on cut flowers (see following sections). The end result of low temperatures, therefore, is longer flower life.

Keeping cut flowers at the coolest temperature that the flower will tolerate will maximise their life. Good growers, wholesalers and florists are always careful to keep cut flowers cool. After picking, growers place their flowers promptly in coolrooms, and transport them in refrigerated trucks. Wholesalers also keep their flowers in cool rooms and transport in refrigerated trucks.

It is important that this 'cool chain' is maintained by the florist. Every florist should have at least one display refrigerator. A small walk-in coolroom is very useful for keeping flowers not on display, or for finished arrangements prior to delivery.

Display your flowers in the coolest part of your shop—don't display outside in hot weather. If outside displays are necessary, only use those flowers that tolerate high temperatures and at least make sure the flowers are kept in the shade.

When you sell flowers, advise purchasers on temperature management:
- take flowers home as quickly as possible: don't leave flowers in hot cars;
- place flowers as soon as possible in a clean vase filled with *cold* water and treated with a preservative; and
- display flowers away from direct sunlight and other sources of heat, such as heaters and lamps.

TROPICAL FLOWERS

Not all flowers, however, need to be cooled. Tropical varieties (see below) are harmed by refrigeration. These flowers grow naturally in tropical regions where the air temperature does not fall below 18°C.

COOLING

It is vital, therefore, to know the appropriate storage temperature for each flower variety.

There are three important temperature ranges:

Full refrigeration (1°C to 4°C)
Most flowers last longest when kept between 1°C and 4°C. Many popular varieties are included in this group, namely Rose, Carnation, Chrysanthemum, all types of Lily, Iris, Tulip, Gypsophila and most types of foliage. Commercial coolrooms and high humidity refrigerators are commonly set around 4°C.

Moderate refrigeration (4°C to 6°C)
Some flowers need to be cooled, but they can be injured by temperatures below 4°C. Flowers in this class include Gladioli, Gerbera, Lisianthus, Alstroemeria, Stephanotis and Anemone.

No refrigeration (above 12°C)
Tropical flowers are damaged by cool temperatures. Flowers in this category include all Orchid varieties, Strelitzia (Bird of Paradise), Heliconia, Anthurium, Ginger, Gloriosa Lily, Poinsettia and Euphorbia. These varieties must never be refrigerated, and should be kept above 12°C at all times.

HUMIDITY

Most flowers should be kept in air with a high relative humidity (90 to 95%) to prevent them from drying out. Humidity refers to the amount of water in the air. Humid days are those when the air contains high levels of water (over about 80% relative humidity), while on hot, dry days the humidity level can fall to as low as 10%. In air-conditioned rooms and offices the air is quite dry, usually about 65% relative humidity. Flowers lose water much faster when the air is dry (low humidity), compared to when the air is damp.

Humidity control can really only be achieved when flowers are in coolers. The ideal humidity for flowers is between 90% to 95%, and it is not possible to maintain this humidity level in the shop, even with regular misting.

Some florists maintain a higher humidity in their display area by misting flowers. This is very helpful with some flowers, such as Hydrangeas, Boronia, Thryptomene and Violets, as they are able to absorb water through their petals and leaves. However, flowers such as Roses, Gerberas and Chrysanthemums are very susceptible to fungal diseases such as *Botrytis*, and misting increases the likelihood of fungal attack. If you are thinking of trying misting, check Chapter 5 (A to Z care guide) for each flower to see if it is recommended.

2. WATER

Water is essential for fresh cut flowers. Some cut flowers, such as Stirlingia, and some gum foliage are sold dried and dyed, and do not need water, but the vast majority of cut flowers are fresh and need a constant supply of clean water.

Cut flowers take up water from the vase through the cut stem end, with the water moving up the stems through very fine tubes called xylem vessels. These vessels are used to conduct water from the roots when flowers are still attached to the plant.

Vase water replaces the water in flowers that is continuously lost by evaporation, which occurs mostly through thin petals, the leaves and stems. If the evaporated water is not replaced at the same rate that it is lost through evaporation, flowers wilt. As we discovered above, the amount of water lost from flowers is determined by the air temperature and humidity; the warmer and drier the air, the more water is evaporated from the flower. For most flowers, 'warm air' means anything over 4°C, so any flower out of the cooler will lose water through evaporation. It is very important, therefore, to make sure water can move up cut flower stems as easily and quickly as possible.

This is not always a simple task. The small xylem vessels in the stem end can easily become plugged by microbes or dirt particles in the vase water. When plugging occurs, water cannot flow easily up to the flower, which can then quickly wilt. Bent neck in Roses is a classic example of stem end plugging.

To prevent plugging, vase water must be *clean*. Vase water that is not changed regularly (at least every second day) or treated with a flower preservative will quickly become a 'soup' as bacteria and fungi multiply and foul the water. Microscopic bacteria and fungi are present even in the cleanest water and buckets. These can only be controlled by adding a germicide, such as chlorine, to the water, and by washing all buckets and vases in a dilute chlorine solution.

Chlorine, which is contained in most household bleaches, is very effective in killing bacteria and fungi in buckets and vases. Low chlorine concentrations (between 10 and 30 parts per million) keep water clean and will not harm flowers.

When using bleach, check that the brand used contains around 4% chlorine. At this concentration, use 15 drops (about ¼ teaspoon) per litre of water to make an effective germicide (see 'Making a chlorine solution' on page 6). Some generic brands of bleach do not specify the chlorine content, which may be as little as 1%. If this is the case, use 1 teaspoon of bleach per litre of water.

An alternative germicide is swimming pool chlorine, which comes in two forms: sodium hypochlorite and the longer lasting sodium dichloroisocyanurate (SDIC). Sodium hypochlorite lasts for two days in vase water, but SDIC is more stable and lasts up to five days. Florists using powdered pool chlorine should use it at the rate of 0.4 g or 2 pinches per 5 L of water. *Note that even water treated with chlorine needs to be changed regularly. It should be replaced with clean water and fresh chlorine at least every three days.* **Be careful!** Chlorine is corrosive. Inhaling chlorine fumes is very damaging to the lungs, so make sure you wear gloves and keep your face averted when handling all forms of chlorine.

Another excellent germicide is bromine. Bromine is less soluble in water than chlorine but can be used as a tablet or with a commercial applicator. Bromine is very effective in low concentrations (10 ppm) and will last for up to five days in solution.

Recent research has shown that chlorine germicides may form carcinogenic compounds in vase water, but this is still a contentious issue. It appears that bromine does not form these potentially harmful compounds.

It is pointless to spend precious time treating vase water if you don't keep all your flower containers and preparation areas clean as well. Buckets, containers and vases should be cleaned following the use of a germicide solution. It is also important to keep your work area and utensils clean. Bacteria and fungi will multiply rapidly in an unclean work area or if you are using unclean equipment. They will then stick to flower stems and pollute vase water. Benches, secateurs, snips and shears should be washed with a germicide solution at the end of each day. Germicidal kitchen sprays are ideal for this.

Some flowers are particularly prone to wilting in dirty water for reasons we do not yet fully understand. Flowers grown in the field, such as Stock, Aster, Dahlia, Gypsophila, Daisy, Candy Tuft, Statice, and Ageratum, seem to have 'dirtier' stems and leaves and will foul vase water quickly if leaves are not stripped and the water is not treated with a germicide. Roses and Gerberas are very sensitive to dirty water, while Tulips, Iris and Carnations are pretty tolerant.

MAKING A CHLORINE SOLUTION
VASE WATER

The ideal chlorine concentration in vase water is between 30 and 50 parts per million. If you use bleach, add ¼ teaspoon per litre if the bleach is a 4% solution (read the label); and 1 teaspoon if the bleach is a 1% solution (most 'no-name' varieties). If you are using swimming pool chlorine, add 1 pinch (0.1 g) per litre.

WASHING BUCKETS

The ideal chlorine concentration for washing buckets and vases is between 10 and 30 parts per million. If you use bleach, add 1 to 2 teaspoons per 10 L if the bleach is a 4% solution (read the label); and 4 to 8 teaspoons if the bleach is a 1% solution (most 'no-name' varieties). If you are using swimming pool chlorine, add 10 pinches (0.8 g) per 10 L. First wash buckets in detergent, then rinse in chlorine solution.

RECUT STEMS

Whenever a flower stem is cut in air, tiny air bubbles are instantly drawn up the xylem vessels, causing a blockage. These bubbles (also known as emboli) become firmly established in most flower stems after about an hour out of water. The longer flowers are kept dry, the more difficult it becomes for water to move past these bubbles.

Thankfully, these air bubbles only move 2 to 3 cm up most flower stems. So by simply cutting 2 to 3 cm off each stem end, it will get rid of these pesky bubbles.

Ideally, recutting stems should be done underwater, as this will prevent more bubbles being formed in the freshly cut stems. Research in the USA has shown that wilted Roses in particular, along with Chrysanthemum, Gladiolus and Snapdragons, recover very well when stems are recut underwater. As a general rule of thumb, any flowers that have been out of water for 30 minutes or more should have at least 2 cm cut from the base of the stems to remove these bubbles.

The simplest way of recutting stems underwater is to use *sharp* secateurs or snips with the stems held about 4 cm underwater in a wide mouthed container or a deep tray. Once stems have been recut underwater they may be held out of water for a few minutes while the flowers are transferred to other vases or placed in floral foam.

Recutting stems underwater can be difficult, and messy, so a good compromise is to cut stems in air then immediately place them in clean water. Remember to always cut flower stems with a sharp knife, scissors or secateurs. Blunt tools damage the ends of flower stems and restrict water uptake. It's also a great idea to cut the stems at an angle so that they will not sit flat on the bottom of the vase, preventing water uptake.

CUTTING METHODS

1 Blunt scissors — compression of edge of stem.

2 Bashing with hammer — all tissues destroyed, leaving large area of damaged tissue that can be infected.

3 Sharp knife — stem able to absorb water.

The practice of smashing woody stems or splitting stems vertically is not recommended. It severely damages the stems and water-conducting vessels and prevents the flower from taking up water. Old theories advocating breaking or hammering stems were based on the principle that it opens up the structural wall of cells to allow rapid water absorption. While this does occur, it only works on a temporary basis. It also severely damages the xylem vessels and makes them a focus for infection from bacteria, and causes small bits of the stem to clog the xylem vessels. Uptake will only be effective if healthy, undamaged cells are present.

WATER TEMPERATURE

All cut flowers should be placed as quickly as possible in cold or room-temperature water. Cold water is particularly recommended as it helps to keep the flowers cool and slows down the opening of buds and wilting of open flowers. Using cold water is a terrific idea in hot weather, as it has the same effect on flowers as a cool drink has on you!

There are times, however, when warm water can also be of benefit, particularly with tightly closed or wilted flowers. Warm water should only be used with very dehydrated flowers, or flowers you want to open quickly. Warm water (30 to 40°C) moves more rapidly up stems than water at room temperature (18°C). Unless they are in very tight bud, most flowers (for example Roses and Tulips) will open too quickly and 'blow' if placed in warm water, and are best placed in cool water. This is also true of winter bulb flowers such as Tulips, Daffodils, Jonquils and Hyacinth, as well as Anemones.

A good example of a flower that can benefit from warm water treatment is Dutch Iris. In mid-winter (July) it can sometimes be difficult to open Iris that are picked in tight bud. A good way of rapidly opening tight Iris is to place them in warm water for about 30 minutes. Keep a close eye on them as they can sometimes open in 5 minutes or less. This is a great way of opening a number of flower types in tight bud. But remember: only place in warm water for a short time (up to 30 minutes) and keep a close eye on them as things can happen fast!

STRIP LEAVES THAT WOULD BE UNDERWATER

Flowers, leaves and stems are naturally covered with bacteria and fungi. Whether they are grown in glasshouses or outside, bacteria and other micro-organisms reside there. These bacteria and fungi will proliferate in water and block the flower stems if left uncontrolled. Field-grown flowers tend to foul vase water more quickly than those grown indoors, which suggests that they may have more bacteria and fungi on them. Typical field-grown flowers are Calendula, Stock, Field Carnations, Daisies, Ageratum and most low-priced bunched flowers.

Regardless of what flower you are using, you must always remove any leaves that would be underwater. By doing this you remove an important source of vase water contamination. It is not necessary, however, to strip thorns from Roses, as this only damages the stems and can lead to disease.

ACIDIFY WATER

Acidifying vase water to a pH of around 3.5 has two advantages: it allows flowers to take up more water, and it also prevents the growth of bacteria. Acidic solutions move more readily through the stems of cut flowers than those that are neutral (such as tap water) or alkaline.

Many commercial flower preservatives contain an acidifier in one form or another, usually citric acid or alum. If you are making your own vase solution, add 5 mL (1 teaspoon) of vinegar or 0.3 g (3 pinches) of citric acid per litre of water.

FLUORIDE SENSITIVITY

A few flowers, notably Gerberas and Gladioli, are very sensitive to the fluoride that is present in most metropolitan tap water. Other sensitive flowers include Freesia, Alstroemeria, Tulip, Snapdragon, Chrysanthemum, Christmas and Asiatic Lilies, and certain Rose varieties. Fluoride damage causes a brownish discolouration of flowers, especially in the outer petals and petal tips. Damage is also evident as brown spots on petals, leaves and stems.

Fluoride is added to tap water in most Australian states to reduce the level of dental decay in the community. Normal concentrations (approximately 1 ppm) in our drinking water are damaging only to some varieties of Gerbera, Gladioli and possibly Freesias.

There is considerable variation in sensitivity to fluoride between Gerbera varieties, with white, pink and yellow varieties being the most sensitive. In Gladioli, red and orange varieties seem to be the most sensitive, whereas yellow and white varieties are partially resistant. An increasing number of fluoride-tolerant varieties are now available and it is rare to see fluoride-damaged flowers these days.

Avoiding tap water is the best way to prevent fluoride damage, but for most florists this is not practical. If you think your flowers have been damaged by fluoride, you should check that your supplier has used fluoride-free water after harvest, as subsequent use of tap water may not cause too much damage. The use of preservatives in vase solutions is also effective in reducing fluoride damage. If you are sure the problem is caused by fluoride, the only advice is to switch to the newer, fluoride-resistant varieties.

3. FLOWER FOOD

When flowers are still attached to the plant, they are provided with all the food they need by the leaves. Flower food is sugar, and is provided to flowers in the form of carbohydrates produced in the leaves by a process called *photosynthesis*. Photosynthesis is the process where plants form sugars from water, carbon dioxide (from the air) and sunlight.

Cutting flowers from the plant separates them from this natural food supply, as well as from water. The good news is that you can replace this food source by adding a flower preservative to vase water. Replacing this food will help extend flower life and, in many cases, can cause buds to develop into mature flowers.

For the vast majority of cut flowers, ordinary cane sugar (sucrose) is the best type of flower food. Commercial flower preservatives contain sugar at a concentration of between 1 and 2%, which is the best level for most cut flowers. Using a flower preservative in the shop and encouraging your customers to use it as well by supplying a sachet, is the best thing you can do to make sure your flowers last as long as possible.

You can make your own preservative if you like, but there are two rules to follow.

Rule 1: Never add sugar to vase water without a germicide. Bacteria feed on sugar, and there is nothing they breed faster in than a nice sugar solution. Placing flowers in a solution containing sugar and no germicide can kill them very quickly.

Rule 2: Always use an acidifier, such as citric acid or vinegar. Sugar solutions are sticky, and flowers cannot take up as much sugar solution as water alone. Adding an acidifier helps the flower take up more of the solution and ensures the sugar gets into the flower to do its work.

Two distinct forms of sugar solution are used to extend the vase life of cut flowers.

Pulsing (or opening) solution: This is a solution with a high sugar content, about 10%. It is used on flowers picked in tight bud to supply them with a high concentration of sugar to open the buds. This solution can be used on Gypsophila, Carnations, Chrysanthemum and Gladiolus or any flower that is in tight bud *except* for Rose, as high sugar solutions damage Roses.

Flowers are usually only placed in a pulsing solution overnight (16 hours) or for one day. *Pulsing should be done by the grower* and should only be used by florists who need to open flowers quickly.

Preservative solution: This commonly contains a germicide and a little sugar (about 1%) and is found in most commercial flower preservatives, whether they are crystals or liquid. Adding 10 g of sugar (about 2 teaspoons) per litre of water along with a germicide and an acidifier will help maximise flower vase life. This solution is effective for the majority of cut flowers.

Some florists prefer to make their own formula, which may vary for different species, and for different purposes, while others are happy with ready-made commercial preservatives. Whichever you favour, most preservatives combine a germicide, an acidifier and plant food (sugar). If you want to make your own preservative, try the recipe below.

STANDARD PRESERVATIVE

Quantity per 5 litres	Quantity per 1 litre
0.4 g (4 pinches) swimming pool chlorine or 1¼ tsp. bleach (if 4% solution) or 5 tsp. bleach (if 1% solution)	¼ tsp. bleach (if 4% solution) or 1 tsp. bleach (if 1% solution)
10 tsp. (50 g) sugar	2 tsp. (10 g) sugar
5 tsp. (25 mL) vinegar or 7–8 pinches (1.5 g) citric acid	1 tsp. (5 mL) vinegar or 3 pinches (0.3 g) citric acid

There are a number of commercial brands of floral preservatives available in both professional and domestic quantities. The simplest of these contain sugar and a germicide, but different ones have different additives, such as citric acid and other compounds. Some come in liquid form, others as granulated powders. Most do an excellent job in extending vase life.

I strongly recommend that florists encourage customers to use a preservative by providing a sachet with every purchase. This can mean the difference between a happy customer, whose flowers last, and someone who will not buy from you again.

Some flowers, notably Orchids and Sandersonia, are sold with their stems placed in vials, usually containing a germicide and sugar. Florists should leave the flowers in the vials. Check that the solution is covering the end of the stem each day, and top up if you can. If the stems are in a transparent sachet, leave them until the solution level is low, then remove and place in flower preservative.

4. ETHYLENE

Ethylene is a gas that kills sensitive flowers. Typical symptoms are petal wilting and flower and leaf drop. You can't see it, you can't smell it, but in urban areas it's always there. Some flowers, such as Carnations, Delphinium and Waxflower, are very sensitive to ethylene gas. However, many flowers, such as Chrysanthemum and Gerbera, are not affected by ethylene at all.

Ethylene is a natural plant hormone that is produced as a gas. It comes from a number of sources: car and truck exhausts, damaged or dying flowers and foliage, ripe fruit and cigarette smoke. It is an accelerator of ageing in flowers, and can drastically reduce the life of sensitive varieties. The table on page 14 lists flowers that are affected by ethylene.

Typical symptoms of ethylene damage include petal or flower drop, as commonly seen in Sweet Pea, Delphinium and Larkspur, Agapanthus, Geraldton Waxflower, Waratah and Snap Dragons. Ethylene also causes petal curling (sometimes called sleepiness) and/or premature wilting in Carnations and Sweet William, Lilium species (Oriental and Christmas Lilies in particular), Gypsophila, Gerberas, Gladioli, Anemones and Stock. Orchids are also sensitive and typically show transparent petals when exposed to ethylene. Other flowers, such as Rose, Alstroemeria and Freesia, fail to open properly when ethylene is present.

ETHYLENE-SENSITIVE FLOWERS

Achillea	Freesia	Physostegia
Aconitum	Geraldton Wax	Rose
Agapanthus	Gerbera	Rudbeckia
Alstroemeria	Gladiolus	Scabiosa
Anemone	Godetia	Snapdragon
Aquilegia	Grevillea	Solidago
Asclepias	Gypsophila	Stephanotis
Astilbe	Iris	Stock
Bouvardia	Kniphofia	Sweet Pea
Campanula	Lilac	Sweet William
Carnation	Lilium	Trachelium
Celosia	Lisianthus	Triteleia
Cornflower	Matthiola	Veronica
Delphinium	Nerine	Verticordia
Eremurus	Orchid	Waratah
Eucomis	Phlox	

SOURCES OF ETHYLENE

The major source of ethylene in city and suburban florists is car fumes and dying flowers. At growers' properties, wholesalers' sheds and at central markets, the principal source is dying flowers and ripening fruit. It is estimated that as little as 0.1 parts per million (ppm) ethylene is sufficient to severely damage carnation flowers. Measurements of ethylene levels outside a number of city and suburban florists located on busy roads range from 1 to 10 parts per million, which is 10 to 100 times the level that can severely damage sensitive flowers. Such flowers placed outside shops are, therefore, constantly exposed to dangerous levels of ethylene.

In a highly competitive commercial environment it may be necessary to advertise the presence of a florist shop by placing flowers on the footpath. Florists should be aware, however, of the damage this practice does to ethylene-sensitive flowers.

I recommend only placing flowers that are not sensitive to ethylene on display, such as: Chrysanthemum, Achillea, tropicals (especially

Heliconia), Strelitzia, Banksia, Calla, Dahlia, foliage (especially Eucalyptus), Gerbera, Erica, Iris, Tulip, Kangaroo Paw, Protea, Leucadendron, Leucospermum, Daisy, Nerine, Poppy, Ranunculus, Statice, Sunflower, Thryptomene and Zinnia.

CONTROLLING ETHYLENE

Fortunately, the effects of ethylene can be minimised by using anti-ethylene treatments, by reducing temperature and by eliminating ethylene sources.

Ethylene-sensitive flowers treated with silver thiosulphate (STS) or the gas methylcyclopropane (MCP) will no longer be affected by ethylene and will last much longer. The best time to apply these treatments is as soon as the flowers are picked. In other words, flowers must be treated by the grower: the earlier they are treated the better.

Florists should not use STS or other anti-ethylene treatments. These treatments are difficult to apply and dispose of, and by the time flowers reach a florist shop it is often too late to treat them effectively.

The best thing florists can do is to check that sensitive varieties have been treated by the grower. In addition, it is a great help if florists ensure that flowers are kept away from the major sources of ethylene:
- remove damaged and dying flowers from bunches and from the floor of the shop or cooler;
- do not store or sell flowers near ripening fruit or vegetables;
- do not display ethylene-sensitive flowers on the footpath;
- transport flowers in enclosed, preferably refrigerated, vans;
- don't smoke cigarettes near flowers.

Chapter 2

BUYING AND PREPARING FOR SALE

Apart from floral design, the main task for florists is to ensure that flowers last as long as possible. It is important, therefore, that the chain of flower care be maintained from the moment the flower is picked to the time it is arranged in your customer's home or office.

Whether the flowers are delivered or bought direct from a farm or wholesale market, florists need to learn to recognise flowers that are fresh and that have been correctly treated by growers and wholesalers. Furthermore, florists should not be afraid to ask questions about their treatment.

The first thing a florist needs to learn is how to buy fresh flowers of the best quality. This is not always easy to do. Sometimes it is nearly impossible to work out if that bunch of Roses or Lilies you want was picked yesterday 20 km down the road or five days ago in Africa.

Many florists rely on good long-term relationships with their suppliers to make sure they always get the quality and freshness they expect. Your growers know that you expect the freshest flowers and will deliver accordingly. Developing these relationships is the best way to do business in the long run, but there will always be times when a new product or grower comes on the market and you will need to be discerning.

For this reason, an entry 'What to look for' is provided for each flower described in Chapter 5. This contains specific instructions on what to look for and, more importantly, what to avoid when buying.

In the meantime, here's a quick buying checklist that will help.
• Check petals for brown marks or other signs of disease (eaten petals, 'cobwebs').
• Check that buds are not too green and that they will open. This is especially important in the southern states during winter.
• Flowers should not be too open (blown). Open flowers last for two days at best.

- It is a great idea to shake bunches of flowers that are known to be ethylene sensitive, such as Waxflower, to see if there is any flower or leaf drop. Bunches that are affected by ethylene often drop flowers, petals or leaves. If flowers or leaves drop off when bunches are shaken, it is a sure sign that the bunch has been affected by ethylene and should not be purchased.
- Leaves should be glossy green, with no sign of yellowing or stripes of lighter green or yellow.
- Turn the bunch over and look at the cut stem ends. Cuts should be green in colour, not brown or black.
- Are stems slimy? Don't buy if they are.
- Look inside growers' or wholesalers' buckets. Are they clean? This is most important for Roses and Gerberas. Iris and Tulips, on the other hand, will tolerate a little mud.

Back at the shop

Once you have received your flowers, you need to know what to do when you are preparing them for sale. Chapter 5 contains specific instructions for each flower type. In general, however, the following procedure should be carried out immediately.

- Place flowers in water, even if stems are not recut and water is straight from the tap. This is vital during warm weather.
- Check flowers for any damage and that they are exactly what you ordered. Make a note if there is anything wrong. Don't rely on your memory!
- Separate bunches. Strip leaves and wash stripped stems to remove field dirt. This is very important with field-grown flowers such as Stock, Calendula, Thryptomene, Tulip, Iris and Ageratum.
- Recut stems: remove about 2 cm with a sharp knife or secateurs. Place immediately in cold water that contains a flower preservative.
- Place in a coolroom or in the shop. If coolroom space is limited, put the most expensive flowers (for example Roses or Lilies) in the cooler.
- You can store flowers in water in a coolroom for short periods (less than a week) at the temperatures listed in Chapter 5. *Remember: long storage periods usually result in a shorter flower life*. After storage, recut stems and place in water with a preservative.

Points to remember

- Keep vases and buckets scrupulously clean.
- After vases and buckets have been used, scrub with a small quantity of household bleach, swimming pool chlorine or other germicide (for example nappy soaker). Vases that are difficult to clean can be filled with a mild germicide solution and left for a few hours or until the brown residue disappears. After the vase is emptied there is no need to rinse away the germicide residue.
- Avoid using metal containers as many ingredients in floral preservatives can react with metal. Use plastic or glass instead.
- Hygiene is important: remove diseased or broken foliage and rubbish from under the benches as quickly as possible as they are sources of ethylene.
- Do not keep fruit with flowers in the coolroom.
- Keep flowers out of draughts. A sheet of plastic placed over the flowers in the coolroom will prevent further drying out. This may also be useful in the workroom.
- Use sharp knives or secateurs to cut stems. Blunt shears will cause stem damage and flowers will wilt more quickly.
- Make sure that flowers are properly wrapped when transported. Wrapping or sleeving will help protect flowers against wilting, damage or even cold.

Specialised techniques

What to do

Trim scalded stems

Flower stems that exude a milky sap or latex, such as Dahlia, Poppy and Helleborus, are often scalded in hot water by the grower to prevent sap leakage. There are pros and cons with this treatment, and it doesn't seem to result in any real benefits or disadvantages.

It is important, though, to cut the scalded section off each stem before you arrange flowers. Scalding is an inexact process and often results in damage to stem ends: the stems curl up and form a U-shape, which severely hinders water uptake. By trimming off the scalded stem sections, you remove this damage and flowers can take up water more freely.

I've tried trimming scalded sections from poppies a number of times. Following this the stems took up more water, more buds opened and the flowers lasted longer than those flowers whose scalded stem sections had been left on.

Searing is sometimes recommended for reviving wilted flowers, especially Roses with bent neck. For Roses, it is better to soak the entire bloom (stem and flower) in lukewarm water for a few hours.

Soaking

Many flowers and types of foliage benefit from being completely immersed in water for short periods of time. Most green foliage, for example, loves a good soak for a few hours in cold water before being placed in arrangements. The best way to soak is to submerge in cold water in a bath or basin for a minimum of two hours, or as long as overnight.

This technique should not be used for yellow or grey foliage, or autumn leaves. Yellow foliage turns brown if soaked in cold water. It is better to recut the stems, strip the lower leaves and stand in water. More detail on treating foliage is given for each plant in Chapter 5.

Soaking is also recommended for the flowers of Hydrangea, Lily-of-the-Valley, Boronia, Thryptomene and Violet. These flowers can absorb water through their flower-heads and leaves. I've tried this a number of times and it works, providing:

1 the water is cold; and
2 soaking lasts less than two hours.

What *NOT* to do

Bashing

Not so long ago it was commonly recommended that most woody cut flower stems should be burnt, bashed, split or scalded to aid longevity. Research has shown repeatedly, however, that rather than help water uptake, these techniques damage the water-carrying vessels in the stem. Damaging stem ends increases the chances of bacteria entering and plugging the stems.

Despite any advice you may have read to the contrary, do not bash stems. Similarly, it is sometimes recommended that the woody branch stems of trees and shrubs such as Azaleas, Rhododendrons and Camellias be treated by scraping the bark from the bottom 5 cm and splitting the stem. There is no benefit to be gained from this practice—a flower preservative is the only treatment you need.

Flowers with woody stems are very thirsty, so cut with sharp secateurs, place them in deep water, and top up the water regularly. Warm water may be beneficial in helping to increase the flow of water up woody stems.

BURNING

Some florists prefer burning to scalding for sappy flowers. This involves holding the stem ends over a flame to seal them. This is not recommended as it damages the cut end of the stem, severely hindering water uptake.

Chapter 3

CUSTOMER SERVICE

For florists, it's an unfortunate fact of life that people don't need to buy flowers. Unlike fruit and vegies, which feed us and keep us healthy, flowers are a luxury. People tend to buy flowers with money left over after they've bought the essentials of life, such as food and clothing, and paid their bills.

When the time comes to buy flowers, the public has a range of options: supermarket, market, green grocer, petrol station, roadside seller and, of course, florist. It would be safe to say that most of these retailers will sell flowers more cheaply than florists. So why should anyone buy from a florist?

The answer is simple: *customer service*. Florists offer a higher level of customer service, ranging from the artistry of arranging flowers to the knowledge and advice they can pass on to their customers. This is what the public wants and is the major reason why good florists can compete with cut-price outlets.

This chapter deals with the technical aspects of customer service: how to get flowers to your customers in the best possible condition, and what flower care advice you need to give them.

DELIVERY TIPS

Florists should take the utmost care that their beautiful arrangements reach the customer looking as good as they did when they left the shop. This is often the hardest and most frustrating job!

First, the flowers must be delivered the correct way.
- Use a van that is enclosed, air-conditioned and, ideally, refrigerated.
- Make sure the delivery vehicle is properly sealed so it does not allow exhaust fumes (a major source of ethylene) to enter the van.
- Never allow arrangements to sit in direct sunlight.
- If the customer is not at home, place the flowers in a cool, sheltered position and place a card announcing the delivery in a prominent spot.

CUSTOMER EDUCATION

Second, any advice you can give your customers on flower care is invaluable. Assume they know nothing! For example, with each arrangement you could provide a *Flower Care* card that lists three or four simple tips. Some tips to include are shown below. Feel free to copy them.

Adding a sachet of flower preservative to each bunch is the perfect way to complete customer service. Never forget that it is the level of service you provide that distinguishes you from other flower sellers.

Flower Care card

Take flowers home immediately
Place in water as quickly as possible.

Choose a clean vase
Clean dirty vases by filling them with warm water, adding a few drops of bleach and letting stand for two or three hours.

Add clean water and a preservative
The preservative will keep the water clean, and feed the flowers.

Strip all leaves from the part of the stem that will be underwater
Leaves contain bacteria and fungi that will quickly breed and foul the water, killing the flowers.

Cut stems at an angle
Use sharp snips or secateurs, not blunt scissors, then place straight into water. Cutting at an angle makes sure the stem will not sit flat on the bottom of the vase and prevent water uptake.

Take care when arranging flowers in the vase
Make sure the stems are not jammed into the vase, or that the ends of the stems are not smashed against the bottom of the vase.

Choose the display site carefully
Keep flowers away from direct sunlight and other sources of heat such as heaters, lamps or televisions.

Change the vase water regularly
Vase water should be changed at least every three days. Retrim the stems by cutting 1 to 2 cm off the ends. Arrangements in foam need to be rewatered *every day*.

Remove dying or damaged flowers and leaves
Remove fallen flowers or leaves to prevent fungal infection or rapid wilting due to ethylene.

Chapter 4

FLOWER ARRANGEMENTS

FLORAL FOAM

Floral foam is a basic tool for all florists, and is essential for floral arrangements. Foam is not, however, to every flower's taste. It is generally agreed that most flowers do not last as long when placed in floral foam as they do when standing in water.

In my experience, some flowers are better suited to foam than others. For example, in flowers with softer stems, stem-rotting problems may occur (Gerberas and Tulips) or stems may tear as you put them into the foam (Daffodils, Iris, Tulips and Hyacinths).

In order to minimise problems with foam you must do two things.

First, ensure that all flowers and foliage that are to be placed in foam have been well prepared or conditioned. Stems should have been cut, leaves stripped from the lower part of the stem, and all stems placed in water containing preservative for a few hours prior to arranging in foam.

Second, thoroughly soak the foam in water treated with flower preservative. The best way to soak foam is by floating it on the preservative solution until it sinks by itself. Never push foam underwater as it will create a large air bubble in the centre. This will persist as a dry spot and any flower stem pushed into the centre of the foam will be unable to take up water. Failure to properly soak foam is a common problem.

Once the foam is prepared, place stems deep into the floral foam—the deeper the better as foams dry out from the top. If any stems need to be repositioned, remove the stem entirely and reinsert in the foam. This will ensure firm contact of the stem with the wet foam.

After inserting stems, never pull them out a little way in order to create a small pool of water. Contrary to popular belief, this *does not* occur.

Pulling out an inserted stem by as little as a 2 mm will break contact between stem and foam, preventing water uptake.

Crowding flowers into small pieces of foam also restricts water uptake. Use a generous piece of foam with enough space and volume to ensure there is adequate water for all the flowers in the arrangement.

Another common problem with arrangements in foam is when customers let them dry out. Once foam is dry it does not re-wet adequately, so it is important to top up the container with water, or water containing a preservative solution, at least once a day. Florists should always advise customers of this.

FLOWERS IN ARRANGEMENTS

Here are some general tips for making flowers last in arrangements.

- Don't overcrowd containers—take care that not too many flower stems are jammed into the container and that the ends of the stems are not crushed on the bottom of the vase, preventing water uptake.
- Don't mix fresh Daffodils and Jonquils (*Narcissus*) with other flowers, particularly Roses. *Narcissus* secrete a toxin from their stems that kills other flowers. If you want to mix *Narcissus* with other flowers, place them in a bucket by themselves for a day, throw the water out, wash their stems, then use them with other flowers. Don't recut the *Narcissus* stems, as this will release the toxin.
- Florists sometimes insert a wire up the middle of flower stems to prevent flower-heads from bending. This is commonly done with Tulips and Gerberas. Some florists oppose this practice, considering it as either unnecessary, given proper handling of the flowers, or as harmful, given that bacteria is likely to be present on the wire. Wiring should always be done on the outside of the stem.
- Metal containers are to be avoided, as the metal can react with flower preservatives and decrease water uptake.
- As is emphasised throughout this book, whether the arrangement is in deep water or in floral foam, flowers will keep longer if they are kept cool and in conditions that stop the flowers and foliage from dehydrating. Avoid direct sunlight, heaters, draughts and air-conditioning units.

Chapter 5

A TO Z CARE GUIDE

ACACIA

Botanical name
A range of Acacia species: *A. buxifolia*, *A. baileyana* (Cootamundra), *A. podalyrrifolia* (Mt Morgan), *A. dealbata* (Black), *A. acinacea* (Gold-dust).

Common name
Wattle, Mimosa.

What to look for
Buy branches when less than half of the flowers are open; the rest should be yellow-coloured buds. Yellow buds will open well in the vase, and branches in yellow bud will last much longer, so flowers should not all be out and 'fluffy'. Avoid branches with brown flowers or with branch tips that are dry and curled.

Temperature
1 to 4°C.

Water
Recut at least 2 cm off branches with sharp secateurs. Strip any foliage that would be underwater and place immediately in water. Adding 3 or 4 drops of detergent to a half-filled bucket of water will help buds come out.

Food
Preservative is essential.

Ethylene
Not sensitive.

ACHILLEA

Botanical name

Achillea species e.g. *A. millefolium* (red), *A. filipendulina* (yellow).

Common name

Yarrow.

What to look for

Buy bunches with brightly coloured flower-heads, not faded. Pollen should be visible, and shed when bunches are shaken. Shake bunches and avoid those with flower drop.

Temperature

1 to 4°C.

Water

Recut 2 cm off stems and strip any leaves that would be underwater. Place wilted stems in warm water for less than 2 hours to revive.

Food

Preservative is probably useful.

Ethylene

Sensitive.

Extra tips

Good for drying. Hang bunches upside down in a dry, warm place.

Aconitum

Botanical name

Aconitum carmichaelii.

Common name

Monkshood, Wolf's Bane.

What to look for

Buy bunches when 1 to 3 of the basal flowers are open and flowers are showing a clean blue colour, not too dark purple. Avoid faded, wrinkled flowers.

Temperature

1 to 4°C.

Water

Recut 2 cm off stems and strip any leaves that would be underwater.

Food

Preservative is essential.

Ethylene

Sensitive.

Extra tips

Take care flowers and sap are poisonous. Always wear gloves when handling.

AGAPANTHUS

Botanical name

Agapanthus africanus.

Common name

Agapanthus, African Lily, Lily of the Nile.

What to look for

Buy when flowers have just opened. Shake bunches and avoid those with flower drop.

Temperature

4 to 8°C.

Water

Recut 2 cm off stems, and place immediately in cold water. Replenish water regularly.

Food

Preservative is essential. It will help buds to open and maintain open flowers.

Ethylene

Sensitive.

Extra tips

Remove dead florets, as they are a source of ethylene. Agapanthus rarely receive anti-ethylene treatments, so they are very prone to flower drop, especially in late summer.

Ageratum

Botanical name

Ageratum houstonianum.

Common name

Ageratum, Pussy Foot.

What to look for

Buy when at least a third to a half of the flowers are open. Avoid branches with yellow leaves, wilted tips and faded flowers.

Temperature

1 to 4°C.

Water

Break bunches apart and recut 2 cm off the stems. Strip any leaves that would be underwater and wash stems under the tap. Replenish water daily.

Food

Preservative is essential.

Ethylene

Not sensitive.

Agonis

Botanical name

Agonis flexuosa, A. linearifolia.

Common name

West Australian Tea Tree, Willow Myrtle.

What to look for

Buy branches with more than half of the flowers open as flowers do not open once picked. Sheds flowers easily, especially if picked when immature, so avoid branches with flowers in bud.

Temperature

1 to 4°C.

Water

Recut 2 cm off branches with sharp secateurs. Strip any leaves that would be underwater. Top up the vase water regularly.

Food

Preservative.

Ethylene

May be sensitive.

ALLIUM

Botanical name

Allium sphaerocephalon (Drumsticks), *A. giganteum* (with round purple flower-heads up to 15 cm in diameter), *A. christophii* (spiky purple), *A. schubertii* (huge pale spidery flower up to 50 cm in diameter).

Common name

Onion Flower, Drumsticks.

What to look for

Buy Drumsticks when the first 3 to 4 whorls of flowers at the bottom of each round flower-head are open, which is about half of the flowers. Buy *A. giganteum* when half of the flowers are open. *Allium* flowers will continue to open in the vase.

Temperature

4 to 8°C.

Water

Clean water is essential. Recut 2 cm off stems and place immediately in cold water.

Food

Preservative.

Ethylene

Not sensitive.

Extra tips

Usually only available in summer. They are quite hardy, but keep bunches away from direct sun and warm air as this can cause them to fade.

ALSTRESIA

Botanical name

Alstroemeria X (unknown).

Common name

Alstresia.

What to look for

Buy at a similar stage to that for Alstroemeria i.e. when buds are fully expanded and coloured with at least one flower per floret open. Avoid bunches with yellow leaves; as with Alstroemeria, this is the first sign of ageing.

Temperature

4 to 8°C.

Water

Recut 2 cm off stems. Strip any leaves that would be underwater. Replenish water daily.

Food

Preservative is essential.

Ethylene

Not known to be sensitive.

ALSTROEMERIA

Botanical name

Alstroemeria aurantiaca varieties.

Common name

Alstroemeria, Peruvian Lily.

What to look for

Bunches should be treated by the grower with an anti-ethylene and anti-leaf-yellowing preservative. Buy when buds are fully expanded and coloured, with at least one flower per floret open. There should be 7–10 florets per stem (depending on variety). Avoid bunches with yellow leaves as this is the first sign of ageing.

Temperature

4 to 8°C.

Water

Recut 2 cm off stems with sharp secateurs, strip any leaves that would be underwater.

Food

Preservative is essential.

Ethylene

Sensitive.

Extra tips

Yellowing leaves are usually the first sign of ageing. Alstroemeria is sensitive to fluoride—ideally use rainwater or deionised water. Wear gloves when handling as it may cause skin reactions.

AMARANTHUS

Botanical name

Amaranthus caudatus, A. hypochondriacus.

Common name

Amaranthus, Love-Lies-Bleeding, Tassel Flower.

What to look for

The best stage to buy Amaranthus is when at least three-quarters of the flowers on the spike are open, as flowers will not open in the vase. Avoid spikes where the tip has hooked over, or the colour has faded to a dull brown colour these are old. Don't be concerned with the small yellow flecks as these are only pollen.

Temperature

4 to 8°C.

Water

Bunches do not normally contain foliage, but strip any traces of leaves if they are present. Recut 2 to 3 cm off each stem and place in cool water with a preservative.

Food

Preservative is essential. Amaranthus will benefit from a preservative as the sugar will help keep the hundreds of small flowers fresh.

Ethylene

Not sensitive.

Amaryllis

Botanical name

Amaryllis belladonna.

Common name

Amaryllis, Belladonna Lily, Naked Lady.

What to look for

Buy when flower buds are fully developed, showing colour or starting to open. Fully open flowers will not last.

Temperature

1 to 4°C.

Water

Recut 2 to 3 cm off each stem and place in cool water with a preservative.

Food

Preservative is recommended.

Ethylene

Not sensitive.

ANEMONE

Botanical name

Anemone coronaria, A. japonica.

Common name

Anemone, Poppy Anemone, Windflower.

What to look for

Buy when buds are fully expanded and coloured and the petals have separated from the centre. Leaves should be glossy and evenly green. Avoid bunches with yellow leaves or overly twisted stems.

Temperature

4 to 8°C.

Water

Break up bunches and recut 2 cm off stems with sharp secateurs. Strip any leaves that would be underwater and wash stems under the tap. Place immediately in cold water.

Food

Preservative.

Ethylene

Very sensitive.

Extra tips

Do not place in the same container as freshly cut Daffodils or Jonquils.

ANTHURIUM

Botanical name

Anthurium andreanum.

Common name

Anthurium, Flamingo Flower.

What to look for

Avoid flowers that have purple or brown/black areas as this indicates chilling injury. The spathe (large flat 'petal') should be glossy, with no spots and/or creases and tears; the spadix (the centre spike, which contains the real flowers) should have more than one-third of the tip smooth i.e. the bottom half to two-thirds looks rough.

Temperature

Above 12°C. *Do not refrigerate.*

Water

Recut 2 cm off stems with sharp secateurs.

Food

Preservative is recommended but not essential.

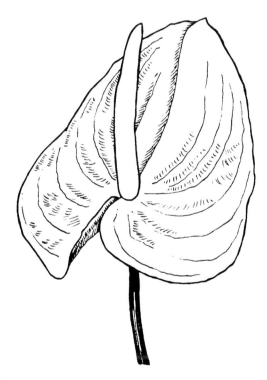

Ethylene

Not sensitive.

Extra tips

Regular misting is recommended. To revive wilted flowers, immerse the whole flower in room-temperature water for 10–30 minutes. Dipping in commercial fruit waxes is said to double vase life.

AQUILEGIA

Botanical name

Aquilegia hybrida.

Common name

Columbine.

What to look for

Each stem should have one open flower, with the rest in bud but showing strong colour. Shake bunches with open flowers, and avoid those with flower drop.

Temperature

1 to 4°C.

Water

Recut 2 cm off stems and strip any leaves that would be underwater. Place immediately into cold water.

Food

Preservative is essential.

Ethylene

Sensitive.

Asclepias

Botanical name
Asclepias tuberosa.

Common name
Butterfly Weed.

What to look for
Buy when a half to two-thirds of the flowers are open, as buds do not open in the vase. Avoid bunches dropping flowers and/or yellow foliage.

Temperature
1 to 4°C.

Water
Place in water treated with a preservative as soon as possible, after stripping any leaves that would be underwater and recutting 2 to 3 cm off each stem. Replace with fresh water daily.

Food
Preservative is essential.

Ethylene
Sensitive.

ASTER

Botanical name

Primarily *Aster novi-belgii*, also *A. ericoides*.

Common name

Aster, China Aster, Easter Daisy.

What to look for

Buy fully opened flowers with unblemished petals. Avoid bunches with yellow foliage. Easter Daisy should be bought when 2 to 4 flowers in each inflorescence (group of flowers) are open.

Temperature

1 to 4°C.

Water

Place in water as soon as possible. Break up bunches, and recut 2 cm off stems with sharp secateurs. Strip any leaves that would be underwater and wash stems under the tap. Asters wilt easily so keep stems in water whenever possible.

Food

Preservative.

Ethylene

Not sensitive.

Extra tips

Asters can develop a strong odour in the vase, so change water regularly.

Astilbe

Botanical name

Astilbe X *arendsii*.

Common name

Astilbe, False Spirea.

What to look for

Astilbe should be given an anti-ethylene treatment by the grower. Buy bunches with a half to a third of flowers open and upper buds coloured. Buds do not develop much after harvest. Avoid bunches with yellow leaves.

Temperature

1 to 4°C.

Water

Break up bunches, and recut 2 cm off stems with sharp secateurs. Strip any leaves that would be underwater.

Food

Preservative.

Ethylene

Very sensitive.

Extra tips

Remove dead leaves, which tend to die off before the flowers. Place wilted bunches in warm water for no longer than 2 hours.

ASTRANTIA

Botanical name

Astrantia major.

Common name

Astrantia.

What to look for

Buy when the upper flowers are fully open. Avoid bunches with all flowers in bud as buds will not open.

Temperature

4 to 8°C.

Water

Recut 2 to 3 cm off each stem.

Food

Preservative is essential.

Ethylene

Not sensitive.

Banksia

Botanical names

A wide range of *Banksia* species are used as cut flowers, including *Banksia coccinea, B. prionotes, B. speciosa, B. menziesii, B. baxteri, B. occidentalis, B. ashbyi, B. hookeriana, B. collina, B. ericifolia, B. spinulosa, B. burdettii.*

Common names

Scarlet Banksia, Golden Banksia, Rickrack Banksia, Raspberry Frost, Lime Green Banksia, Red Banksia, Ashby's Banksia, Hooker's Banksia, Hairpin Banksia, Golden Candlesticks, Burdett's Banksia.

What to look for

Buy when the tiny flowers in the collar around the base of the flower-head are open. The bottom third of the flower-head should be open and 'fluffy'. Colour should be clear; avoid flowers with a grey tinge. Always avoid wet flowers—Banksias picked and packed wet develop black streaks, especially the apricot-coloured varieties.

Temperature

1 to 4°C.

Water

Recut 2 to 3 cm off stems with sharp secateurs, strip any leaves that would be underwater. Never bash or split stems. Place immediately in cold water.

Food

Preservative is optional.

Ethylene

Not sensitive.

Extra tips

Do not mist.

BEARDED IRIS

Botanical name

Iris X germanica.

Common name

Bearded Iris, Flag Iris.

What to look for

The ideal time to buy Bearded Iris is when the buds are showing colour but are not open.

Temperature

1 to 4°C.

Water

Recut 1 to 2 cm off each stem and place in cool water treated with a flower preservative. Keep in a cool place, away from direct sunlight and warm air.

Food

Preservative is essential. The preservative will help buds develop and ensure colours are rich and deep.

Ethylene

Not sensitive.

Extra tips

Each Bearded Iris flower will last about 4 days but each bud can provide up to 3 flowers each. By pulling off each spent flower, a new flower will develop, adding to the vase life of each stem considerably.

Beaufortia

Botanical name

Beaufortia sparsa.

Common name

Sparsa, Swamp Bottlebrush.

What to look for

Buy branches with half-open flowers. Buds where the red 'fluff' has appeared will continue to open, but tight buds may not. Don't buy branches with wilted and faded flowers.

Temperature

1 to 4°C.

Water

Recut 1 to 2 cm off each stem with sharp secateurs, strip any leaves that would be underwater and place stems in cool water.

Food

Preservative is essential. The preservative will help buds develop and ensure open flowers last.

Ethylene

Not sensitive.

BERRIES

Botanical name
A range of species, usually available in autumn, such as *Sorbus, Cotoneaster, Pyracantha; Viscum alba* (Mistletoe) at Christmas.

Common name
Mistletoe, Cotoneaster, Hawthorn.

What to look for
Buy branches with full plump berries that show no sign of shrivelling. Shake bunches and avoid those with berry or leaf drop.

Temperature
1 to 4°C.

Water
Recut 2 cm off stems with sharp secateurs and strip any leaves that would be underwater. Place in cold water immediately.

Food
Preservative.

Ethylene
May be sensitive. Mistletoe is very sensitive.

Extra tips
Some florists recommend stripping all the leaves off to allow berries to be seen. Spraying with hair spray is very useful to prevent berries from shrivelling too quickly.

BIRD OF PARADISE

Botanical name

Strelitzia reginae.

Common name

Bird of Paradise.

What to look for

The first floret in the flower should be open i.e. poking up out of the split sheath surrounding the flower. Avoid flowers with brown marks as these are a sign of chilling injury, which is very common in the southern states during winter.

Temperature

Above 12°C. *Do not refrigerate.*

Water

Recut 2 cm off stems of both the flowers and leaves (if they are included with the bunch) with sharp secateurs. Placing in warm water will help bring the flower out. Replenish water frequently.

Food

Preservative is recommended but not essential.

Ethylene

Not sensitive.

Extra tips

There are several florets in each Strelitzia flower. To bring out these extra florets soak the entire flower in lukewarm water for 20 minutes, then carefully insert your thumb inside the sheath through the slit on the upper side, and gently lift up the new floret. Move the thin white membrane that separates each flower and cut it off.

BLOSSOM

Botanical name

Range of species *Prunus, Chaenomeles* (Japonica or Flowering Quince), *Rhododendron, Azalea, Camellia.*

Common name

Blossom, Flowering Cherry, Flowering Crabapple etc.

What to look for

Buy when branches are in bud, with colour showing, or as partially open flowers. Buds will open after harvest. Shake Prunus branches and avoid those with petal drop, which is common late in spring. Buy Rhododendron when the buds have cracked open and are clearly showing colour. Avoid branches with all flowers fully open as these last about half as long.

Temperature

1 to 4°C.

Water

Recut 2 cm off branches with sharp secateurs and strip any leaves that would be underwater. Place immediately in cold water. Do not bash stems.

Food

Preservative is essential. It is especially important early in the season as it helps buds to open.

Ethylene

Not sensitive.

Extra tips

Some flowering shrubs produce very delicate flowers that are easily bruised and have a short vase life. These should be handled with great care.

Boronia

Botanical name

Boronia heterophylla, B. megastigma.

Common names

Pink Boronia, Lipstick, Brown Boronia.

What to look for

Avoid bunches with wilted tips, yellow foliage or signs of leaf or petal drop. For strong-smelling varieties (e.g. Brown Boronia), buy bunches with strong scent, as this is the best indicator of freshness.

Temperature

1 to 4°C.

Water

Break open bunches and recut 2 cm off stems. Strip any foliage that would be underwater. Wash stems under the tap then place in cold water.

Food

Preservative.

Ethylene

B. heterophylla may be sensitive.

Extra tips

Likes to be misted. Handle carefully as florets drop easily. Wilted bunches can be revived by immersing the entire stem underwater for 2 to 3 hours.

BOUVARDIA

Botanical name

Bouvardia leiantha hybrids, *B. humboldtii* (perfumed white).

Common name

Bouvardia.

What to look for

Buy bunches with open flowers showing bright colours and no brown marks. Leaves should be bright green with no brown tips.

Temperature

4 to 8°C.

Water

Recut 2 cm off stems with sharp secateurs. Strip any leaves that would be underwater. Replenish water frequently.

Food

Preservative.

Ethylene

Very sensitive.

Extra tips

Very prone to wilting. Keep cool, away from hot, dry environments, direct sunlight and draughts.

Calendula

Botanical name

Calendula officinalis.

Common name

Calendula, English Marigold.

What to look for

Buy when flowers are open or half-open. Leaves should be free of yellow blotches.

Temperature

1 to 4°C.

Water

Break open bunches and recut 2 cm off stems. Strip any leaves that would be underwater. Wash stems under the tap, then place bunches in cold water.

Food

Preservative is essential. It will help maintain open flowers.

Ethylene

Not sensitive.

Callistemon

Botanical name

Callistemon species e.g. *C. viminalis* varieties, *C. citrinus* varieties.

Common name

Bottlebrush.

What to look for

Buy when flowers are half-open and looking 'fluffy'. Avoid wilted flowers or branches where most flowers are in tight bud.

Temperature

1 to 4°C.

Water

Recut 2 cm off stems and strip any leaves that would be underwater. Place in cold water.

Food

Preservative is essential. It will help buds come out and maintain open flowers.

Ethylene

Not sensitive.

Extra tips

Flowers are made up of long stamens that wilt readily. Recut stems daily and place in fresh preservative.

CAMPANULA

Botanical name

Campanula species e.g. *C. medium* (Canterbury Bell).

Common name

Canterbury Bell, Bellflower, Bluebell.

What to look for

Buy when a third to a half of the flowers on the bottom of the spike are open. Buy Canterbury Bells when one to two bells per stem are fully open, and the buds are swollen and showing colour. Avoid spikes with yellow leaves and wilted tips. Shake bunches and avoid those where flowers drop.

Temperature

1 to 4°C.

Water

Recut 2 cm off stems. Strip leaves that would be underwater. Leaves will turn yellow before the flowers wilt, so remove as many lower leaves as possible. Replenish water daily.

Food

Preservative is essential.

Ethylene

Sensitive.

CARNATION

Botanical name

Dianthus caryophyllus (Sim and spray), D. chinensis (Chinnies), D. barbatus (Sweet William).

Common name

Carnation, Sim, standard, spray, Chinnies, Sweet William.

What to look for

Buy bunches with clean, undamaged petals that are not curling inwards, known as sleepiness. Sims and sprays are usually sold half-open, and will open in the vase; Chinnies and Sweet William are sold more open. Buy Sweet William when about a quarter to a half of the flowers are open. Carnations are very prone to fungal diseases. Avoid bunches with brown tips on leaves, or brown markings on the petals, which are signs of fungal attack. Bunches must be given an anti-ethylene treatment by the grower.

Temperature

1 to 4°C.

Water

Recut 2 cm off stems of flowers and leaves with sharp secateurs. Make the cut just above a node, which is a knobbly bit of the stem. Strip any leaves that would be underwater. Replenish water frequently.

Food

Preservative.

Ethylene

Very sensitive.

Carthamus

Botanical name

Carthamus tinctorius.

Common name

Matches, Safflower.

What to look for

Buy when the orange, thistle-like flowers are starting to open, with colour clearly visible. Buds do not open in the vase. Leaves should be dark green, with no sign of yellow.

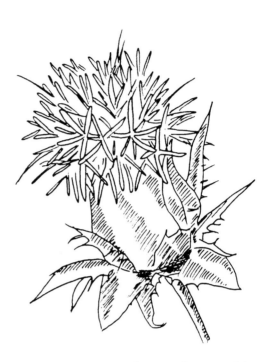

Temperature

1 to 4°C.

Water

Recut 2 cm off stems with sharp secateurs. Strip any leaves that would be underwater as leaves go limp and rot quickly. Leaves tend to yellow before flowers wilt, so you may want to strip all leaves. Change water daily.

Food

Preservative is essential.

Ethylene

May be sensitive.

CELOSIA

Botanical name

Celosia argentea pyramidalis (Prince of Wales Feather), *C. argentea cristata* (Cockscomb), *C. spicata* (Wheat Celosia).

Common name

Prince of Wales Feather, Cockscomb, Wheat Celosia.

What to look for

Buy fully developed flowers; flowers will not open further after they are cut. First look at the leaves, which should have no yellow or brown marks. Flowers should be brightly coloured.

Temperature

1 to 4°C.

Water

Recut 2 cm off stems with sharp secateurs. Strip any leaves that would be underwater as leaves go limp and rot quickly. Change water daily as the thick, leafy stems rapidly pollute the vase solution.

Food

Preservative is essential.

Ethylene

Sensitive.

CHRISTMAS BELLS

Botanical name

Blandfordia grandiflora.

Common name

Christmas Bells.

What to look for

Buy stems with at least one bell open and with fresh, undamaged flowers. Bells are easily crushed, leaving black creases.

Temperature

4 to 8°C.

Water

Recut 2 cm off stems, and place straight into water. Check vase water often as Bells are thirsty drinkers.

Food

Preservative is essential—it will help open buds.

Ethylene

Not sensitive.

CHRISTMAS BUSH

Botanical name

Ceratopetalum gummiferum.

Common name

Christmas Bush, Festival Bush.

What to look for

Look for bunches showing strong red/yellow colour. The 'flowers' should be free of marks, with the 5 calyces flexed back. Avoid dark, purplish flowers, as this is an indication that they have dried out.

Temperature

4 to 8°C.

Water

Recut 2 cm off stems with sharp secateurs, and strip any leaves that would be underwater. Check vase water often as they are thirsty drinkers.

Food

Preservative is essential.

Ethylene

Not sensitive.

Extra tips

Regular misting is highly recommended as branches take up water through the flowers and foliage as well as through the stems.

CHRYSANTHEMUM

Botanical name
Dendranthema species, mainly *D. morifolium*.

Common name
Chrysanthemum, Pompom, Spiders, Disbuds.

What to look for
Buy sprays when the most advanced flower on the spray is fully open
and is beginning to produce pollen. Buy Pompoms when the 'tuft'
appears i.e. when centre petals are no longer sticking together. Look for
clean, undamaged flowers and foliage—yellow foliage is the first sign of
ageing. Make sure the woody base of the stem is not remaining.

Temperature
1 to 4°C.

Water
Recut 2 cm off the
stems with sharp
secateurs, making sure
you remove the
woody basal part of
the stem. Strip any
leaves that would be
underwater as leaves go
limp and rot quickly.

Food
Preservative is essential.

Ethylene
Not sensitive.

Extra tips
Never crush or pound stems.

CLIVEA

Botanical name
Clivea miniata and *C. nobilis*.

Common name
Clivea.

What to look for
Look for clean unmarked flowers. Buy when the first flowers on the cluster have started to open.

Temperature
1 to 4°C.

Water
Recut 2 cm off stems and place immediately in cold water.

Food
Preservative.

Ethylene
Not sensitive.

Conospermum

Botanical names

Conospermum amoenum, C. crassinervium, C. incurvum, C. stoechadis.

Common name

Smokebush.

What to look for

Buy bunches with undamaged flowers and with a fluffy appearance. Flowers of some species are naturally soft and grey, and may appear wilted when they are fresh. Best indication of freshness is the fluffiness of flowers.

Temperature

4 to 8°C.

Water

Recut 2 cm off stems, strip any leaves that would be underwater and place in cold water.

Food

Preservative.

Ethylene

Not sensitive.

Extra tips

Some growers suggest that Smokebush varieties without leaves are best kept dry, as they shed their flowers when placed in water.

COREOPSIS

Botanical name

Coreopsis lanceolata.

Common name

Coreopsis, Tickseed.

What to look for

Buy when flowers are open. Leaves should have no sign of yellowing.

Temperature

1 to 4°C.

Water

Break open bunches and recut 2 cm off stems and strip any leaves that would be underwater. Wash stems under the tap, then place in cold water.

Food

Preservative is essential. It will help maintain open flowers.

Ethylene

Not sensitive.

Extra tips

Often picked from the wild in NSW, where it grows happily along railway lines and roads as a weed.

CORNFLOWER

Botanical name

Centaurea cyanus (blue), C. macrocephala (large, yellow).

Common name

Cornflower, Bachelor's Buttons.

What to look for

Buy when flowers are half-open. Fully open flowers clearly showing flecks of pollen are older and will not last. Check bunches for broken stems. Avoid bunches with yellowing leaves or blackened cut stem ends.

Temperature

1 to 4°C.

Water

Split bunches and recut 2 cm off the stems. Strip leaves that would be underwater and wash stems under the tap. Replenish water daily.

Food

Preservative is essential. It will help maintain open flowers.

Ethylene

Sensitive.

COSMOS

Botanical name

Cosmos bipinnatus.

Common name

Cosmos, Mexican Aster.

What to look for

Buy bunches when the first one or two flowers are open, but petals have not flattened. Flower centres should be yellow but not producing pollen. Avoid bunches with wilted and faded flowers and/or yellow foliage.

Temperature

1 to 4°C.

Water

Break open bunches and cut 1 to 2 cm off each stem with sharp secateurs, strip any leaves that would be underwater and wash stems under the tap. Place immediately in cool water treated with a flower preservative.

Food

Preservative is essential. The preservative will help buds develop and ensure open flowers last.

Ethylene

Not sensitive.

CROCOSMIA

Botanical name
> *Crocosmia* hybrids.

Common name
> Crocosmia, Montbretia.

What to look for
> Buy when the lower buds are showing full colour or the first flowers on the cluster have started to open. Shake bunches—there should be no sign of flower drop.

Temperature
> 4 to 8°C.

Water
> Break open bunches and cut 1 to 2 cm off each stem with sharp secateurs, strip any leaves that would be underwater and wash stems under the tap. Place immediately in cool water treated with a flower preservative.

Food
> Preservative.

Ethylene
> Very sensitive.

Extra tips
> Crocosmia grows as a weed, and is usually harvested from roadsides, so it is most unlikely that bunches are treated against ethylene. They are therefore very prone to flower drop. Bunches with seed pods are available in autumn and these do not appear to drop.

CROWEA

Botanical name

Crowea exalata varieties.

Common name

Crowea.

What to look for

Buy when flowers are half-open or in coloured bud. Flower drop is a problem in old bunches and late in the season (late spring to early summer).

Temperature

1 to 4°C.

Water

Recut 2 cm off stems and strip any leaves that would be underwater.

Food

Preservative is essential. It will help buds come out and maintain open flowers.

Ethylene

Not sensitive.

DAHLIA

Botanical name
Dahlia pinnata.

Common name
Dahlia.

What to look for
Buy when the flowers are three-quarters to fully open but the flower centres are tight. Tight buds will not open. Foliage should be firm and green, with no sign of wilting or yellowing.

Temperature
1 to 4°C.

Water
Break bunches apart and recut 2 cm off stems with sharp secateurs, then wash stems under cold water. Strip any leaves that would be underwater as leaves will rot quickly.

Food
Preservative is essential.

Ethylene
Sensitive.

Extra tips
Remove most of the leaves, as they tend to turn yellow before the flower wilts.

DAISY

Botanical name

Dendranthema frutescens.

Common names

Daisy, Marguerite Daisy.

What to look for

Buy when one or two flowers are completely open and the rest are half-open. Flowers should be perfectly round, with no spaces between petals and no petal damage. Avoid bunches with yellow leaves as this is the first sign of ageing.

Temperature

1 to 4°C.

Water

Recut 2 cm off each stem with sharp secateurs and place in clean water. Remove foliage that would be underwater. Foliage turns yellow before flowers wilt, so strip as many leaves from the bottom of the stem as possible.

Food

Preservative is essential. This will help buds open and delay leaf yellowing.

Ethylene

Not sensitive.

DAPHNE

Botanical name
Daphne odora.

Common name
Daphne.

What to look for
Buy when the small waxy flowers are open. Closed flowers are usually old and not in bud. Leaves should have no sign of yellowing. Check for scent, as this is a sign of freshness.

Temperature
1 to 4°C.

Water
Recut 2 cm off branches with sharp secateurs, and strip any leaves that would be underwater. Place in cold water. Do not bash stems.

Food
Preservative is essential. It will help maintain open flowers.

Ethylene
Not sensitive.

DELPHINIUM AND LARKSPUR

Botanical name

Delphinium grandiflorum; Consolida ambigua and *C. orientalis*.

Common name

Delphinium, Larkspur.

What to look for

Buy when at least a third of the lower flowers on the spike are open. Look for strong stems that fully support the spike. Avoid spikes showing signs of flower drop as this is a sign of ethylene damage. Avoid dry, hooked tips on the spike.

Temperature

1 to 4°C.

Water

Unpack immediately as florets will drop if they are allowed to dry out. Recut 2 cm off each stem with sharp secateurs and strip any leaves that would be underwater as leaves will rot quickly. Place immediately in cold water.

Food

Preservative is essential.

Ethylene

Very sensitive.

Extra tips

Handle with great care as florets drop easily. Wilted florets should be removed as they produce ethylene that will affect the rest of the flower.

DRYANDRA

Botanical name

Dryandra formosa, D. polycephala, D. quercifolia.

Common name

Dryandra, Bush Rose.

What to look for

Buy bunches of *D. formosa* when flowers are Tulip-shaped i.e. when the flowers have swelled but the stamens have not yet popped out. Buy *D. polycephala* when flowers are half-open i.e. looking fluffy. Yellow foliage is normal, but bunches with brown tips on the leaves should be avoided.

Temperature

1 to 4°C.

Water

Recut 1 to 2 cm off each stem with sharp secateurs, strip any leaves that would be underwater and wash stems under the tap. Wear gloves as Dryandra are prickly! Place immediately in cool water treated with a flower preservative.

Food

Preservative is essential. The preservative will help buds develop and ensure open flowers last.

Ethylene

Not sensitive.

ECHINACEA

Botanical name

Echinacea purpurea.

Common name

Coneflower.

What to look for

Buy when the flowers are fully open and petals have flexed back. Sometimes sold with petals removed (see below). Leaves should have no sign of yellowing.

Temperature

1 to 4°C.

Water

Recut 2 cm off stems and strip any leaves that would be underwater. Wash stems under the tap, then place in cold water.

Food

Preservative is essential. It will help maintain open flowers.

Ethylene

Not sensitive.

Extra tips

Petals often wilt after a few days, so it is a good idea to strip these off altogether and use the colourful cone.

ECHINOPS

Botanical name

Echinops bannaticus.

Common name

Globe Thistle.

What to look for

Buy when about half of the globe-like flower-head is blue and open. Avoid bunches with yellow leaves.

Temperature

1 to 4°C.

Water

Place in water treated with a preservative as soon as possible, after stripping any leaves that would be underwater and recutting 2 to 3 cm off each stem. Leaves turn yellow before the flower-head fades, so strip off as many leaves as possible.

Food

Preservative is essential.

Ethylene

Not sensitive.

Eremurus

Botanical name

Eremurus robustus.

Common name

Foxtail Lily, Desert Candle.

What to look for

Buy when the first 2 or 3 flowers on the spike have opened. Look for stems with straight tips pointing upwards as bent spikes will not straighten.

Temperature

Above 12°C. *Do not refrigerate.*

Water

Recut 2 cm off stems. Strip leaves that will be underwater. Replenish water daily.

Food

Preservative is essential to help buds develop.

Ethylene

Sensitive.

Extra tips

Remove wilted flowers as they are a source of ethylene.

Erica

Botanical name
Erica species, including *E. sessilifolia*, *E. hybrida*, *E. baccans*, *E. cerinthoides*.

Common name
Heath, Heather.

What to look for
Choose branches with at least half of the flowers open. Look at the small individual flowers, and buy bunches whose flowers are plump, not shrivelled.

Temperature
1 to 4°C.

Water
Recut 2 cm off each branch with sharp secateurs. Strip any leaves that would be underwater. Change water regularly to avoid bacterial build-up.

Food
Preservative is essential.

Ethylene
Not sensitive.

Extra tips
Regular misting will help to prevent flower drop.

ERIOSTEMON

Botanical name

Eriostemon australasius, E. myoporoides.

Common name

Eriostemon, Eastern Waxflower.

What to look for

Buy bunches when flowers are just starting to open and buds are fully coloured pink or white, depending on variety. Avoid bunches with shrivelled flowers. Leaves should be glossy green with no sign of yellowing. Flower drop becomes a problem late in the season (late spring to summer).

Temperature

1 to 4°C.

Water

Recut 1 to 2 cm off each stem with sharp secateurs, strip any leaves that would be underwater and wash stems under the tap. Place immediately in cool water treated with a flower preservative.

Food

Preservative is essential. The preservative will help buds develop and ensure open flowers last.

Ethylene

Sensitive.

Eryngium

Botanical name

Eryngium planum.

Common name

Sea Holly.

What to look for

Buy when the flowers are a clear blue colour. Avoid bunches with yellow leaves.

Temperature

1 to 4°C.

Water

Place in water treated with a preservative as soon as possible, after stripping any leaves that would be underwater and recutting 2 to 3 cm off each stem.

Food

Preservative is essential.

Ethylene

Not sensitive.

EUPHORBIA

Botanical name
Euphorbia species, usually *E. wulfenii* (pale green), *E. polychroma* (yellow and green), *E. marginata* (white and green).

Common name
Euphorbia, Snow on the Mountain (*E. marginata*).

What to look for
Buy *E. polychroma* when the yellow-green flowers are open. Buy *E. marginata* when the white bracts are fully coloured. Check bunches for broken stems. Leaves should show no sign of yellowing.

Temperature
4 to 8°C.

Water
Break open bunches and recut 2 cm off stems *underwater*, as this will prevent the milky latex from seeping from the stems. Strip any leaves that would be underwater. Foliage turns yellow relatively quickly, so you may want to strip all the leaves. Wash stems under the tap, then place in cold water.

Food
Preservative is essential. It will help maintain open flowers.

Ethylene
Not sensitive.

Extra tips
Euphorbias have a poisonous milky latex that can leak from the stems if they are not cut underwater, so always wear gloves when handling.

EVERLASTING DAISY

Botanical name

Rhodanthe, Waitzia species and *Bracteantha bracteata*.

Common name

Everlasting Daisy.

What to look for

Buy when flowers are half to fully open and buds are showing strong colour. Leaves should be dark green.

Temperature

4 to 8°C.

Water

Recut 2 cm off stems and strip any leaves that would be underwater.

Food

Preservative is optional. Use 30 ppm chlorine in the vase water if not using preservative.

Ethylene

Not sensitive.

Extra tips

These daisies dry exceptionally well, hence their name. To dry, hang bunches upside down in a warm dry room.

FLANNEL FLOWER

Botanical name
Actinotus helianthi.

Common name
Flannel Flower.

What to look for
Buy fresh, undamaged flowers with a fluffy appearance. Avoid bunches with floppy tips or, late in the season, bunches containing seed heads.

Temperature
4 to 8°C.

Water
Recut 2 cm off stems, and strip leaves as they will rot underwater. Change vase water regularly.

Food
Preservative.

Ethylene
Not sensitive.

Extra tips
The fine hairs of these flowers make an invisible dust, which can cause allergies in some people and, in severe cases, a choking sensation.

FLOWERING GUM

Botanical name

Eucalyptus species, usually *E. ficifolia*.

Common name

Flowering Gum.

What to look for

Buy when flowers are half-out and look 'fluffy'. Flowers are made up of long stamens and will wilt rapidly.

Temperature

1 to 4°C.

Water

Recut 2 cm off branches with sharp secateurs and strip any leaves that would be underwater.

Food

Preservative is essential. It will help maintain open flowers.

Ethylene

Not sensitive.

Extra tips

Like other flowers that consist of long stamens rather than petals, Flowering Gum will not last more than 5 days or so.

FOLIAGE—
AUSTRALIAN SPECIES

Botanical names
A wide range of species from the genera *Acacia, Eucalyptus, Grevillea, Persoonia, Caustis, Xanthorrhea, Pittosporum…*

Common names
Wattle, Gum, Grevillea, Snottygobble, Emu Bush, Koala Fern, Steel Grass…

What to look for
Choose foliage with firm, undamaged leaves, and avoid branches with wilted tips. Recut out immature tips if necessary.

Temperature
1 to 4°C.

Water
Recut 2 cm off branches with sharp secateurs, and strip leaves that would be underwater. Change vase water regularly.

Food
Preservative is optional.

Ethylene
Not sensitive.

Extra tips
You can store foliage in water at 1°C for up to 2 weeks.

FOLIAGE—EXOTIC SPECIES

There is a huge and ever-increasing range of exotic foliage available—both tropical and temperate. All foliage can be treated the same way, except for tropical foliage, which should never be refrigerated. Listed below are only the most common types.

Botanical names

Nephrolepis (Fishbone Fern), *Camellia*, *Cupressus* (Cypress), *Asparagus*, *Hedera helix* (Ivy), *Quercus robur* (Oak), *Fagus* species (Beech), *Ulmus* species (Elm).

Common names

Ferns Asparagus, Sperengi, Fishbone;
Evergreen Cypress, Laurel, Camellia, Rhododendron;
Deciduous (autumn) Oak, Elm, Beech, Ivy;
Tropical Cordyline, Croton, Dracaena.

What to look for

Buy fresh-looking leaves. Avoid branches with wilted tips or yellowish leaves (except for autumn leaves, of course). Avoid leaves with brown spots.

Temperature

1 to 4°C, except for tropical leaves, which must be kept above 12°C.

Water

Recut 2 cm off each branch with sharp secateurs. Strip any leaves that would be underwater as leaves will rot quickly. Most foliages are heavy drinkers, so always keep in water and replenish regularly.

Food

Preservative is optional, but always use chlorine. Don't place Beech or Asparagus Fern in solutions containing sugar.

Ethylene

Not sensitive.

Extra tips

Wilted foliage can be revived by soaking in cold water for 2 to 12 hours. Do not soak yellow or grey foliage, or tropicals. Misting foliage is highly recommended.

FORSYTHIA

Botanical name

Forsythia species, usually *F. suspensa*.

Common name

Golden Bell, Forsythia.

What to look for

Branches are usually sold in bud, or when flowers are just opening, in late winter or early spring. Branches with all the flowers open will not last as long.

Temperature

1 to 4°C.

Water

Recut 2 cm off each branch with sharp secateurs. Strip any leaves that would be underwater as leaves will rot quickly. Change water regularly to avoid bacterial build-up.

Food

Preservative is essential.

Ethylene

Not sensitive.

FOXGLOVE

Botanical name

Digitalis purpurea.

Common name

Foxglove.

What to look for

Buy when the lower third of florets on the spike are open or starting to open.

Temperature

1 to 4°C.

Water

Recut 2 cm off stems and strip any leaves that would be underwater. Recut stem ends frequently as they are prone to blockage.

Food

Preservative is essential. Other florets will open in the vase if a preservative is used.

Ethylene

Very sensitive.

Extra tips

Wilted florets should be removed as they produce ethylene, which will affect the rest of the flower.

Freesia

Botanical name
Freesia hybrida.

Common name
Freesia.

What to look for
The first flower in each spike should be fully coloured and starting to open. Avoid tight buds—flowers will rarely open if picked before the first bud has started to open. There should be at least 5 but preferably 7 flowers per spike. Make sure there is no tipburn (the drying out and browning of flower tips), which indicates water stress.

Temperature
1 to 4°C.

Water
Recut 2 cm off each stem with sharp secateurs. Strip any leaves that would be underwater as they will rot quickly. Change water regularly to avoid bacterial build-up.

Food
Preservative is essential.

Ethylene
Sensitive.

Extra tips
Don't arrange with *Narcissus* (Daffodils and Jonquils) as the sap from these flowers kills Freesias. Some Freesia varieties are sensitive to fluoride in tap water, which results in leaf burning, smaller flowers, and the failure of smaller buds to open.

GALTONIA

Botanical name

Galtonia candicans.

Common name

Berg Lily, Giant Snowdrop.

What to look for

Buy when about half the white, bell-shaped flowers are open. Flowers should be pale green to white in colour and be free of brown marks.

Temperature

1 to 4°C.

Water

Recut 2 cm off stems, then place in cold water.

Food

Preservative is essential. It will help maintain open flowers.

Ethylene

Not sensitive.

GARDENIA

Botanical name

Gardenia jasminoides.

Common name

Gardenia, Cape Jasmine.

What to look for

Buy clean, crisp white flowers with no brown marks or transparent-looking petals.

Temperature

1 to 4°C. Keep refrigerated at all times.

Water

Stems are too short to trim. Place on cotton wool soaked in preservative, or float in a shallow tray filled with water. Alternatively, place on wet cotton wool in a sealed plastic container and put in the fridge.

Food

Preservative.

Ethylene

Not sensitive.

Extra tips

Gardenias have a short vase life of 2 to 4 days. The flowers are very delicate and turn brown if touched. Handle carefully by the stems.

GELEZNOWIA

Botanical name

Geleznowia verrucosa.

Common name

Yellow Bells.

What to look for

Buy bunches when the yellow flowers are half- to fully open. Flowers dry out in a few days, so check that they are still fresh. Leaves should be khaki-green with some tinges of yellow. Avoid bunches with brown tips on the leaves.

Temperature

1 to 4°C.

Water

Break open bunches and cut 1 to 2 cm off each stem with sharp secateurs, strip any leaves that would be underwater and wash stems under the tap. Place immediately in cool water treated with a flower preservative.

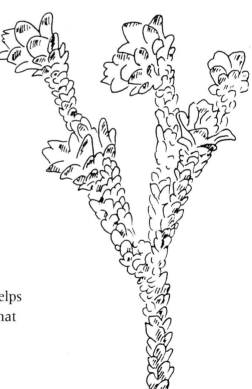

Food

Preservative is essential as it helps buds to develop and ensures that open flowers last.

Ethylene

Not sensitive.

GERALDTON WAX

Botanical name

Usually *Chamelaucium uncinatum*, also *C. megalopetalum*, *C. ciliatum*.

Common name

Geraldton Wax, Waxflower, Bud Wax.

What to look for

Buy branches with full, unopened buds (Bud Wax), or a third to a half of flowers open. Open flowers should be evenly coloured, with no sign of dark areas or petals eaten away, as this indicates fungal infection. Leaves should be green, not yellowish. Branches must be treated with an anti-ethylene agent by the grower. Shake branches, and do not buy when flowers, leaves and buds fall off readily as branches will have been affected irreversibly by ethylene.

Temperature

1 to 4°C.

Water

Recut at least 2 cm off branches with sharp secateurs. Recut underwater if possible and strip any small branches and leaves that would be underwater. Waxflower is very thirsty so top up vase water regularly.

Food

Don't use commercial preservative. Place in clean water treated with chlorine. The sugar in preservatives causes excess nectar production, which can lead to problems with insects and fungal diseases.

Ethylene

Very sensitive.

Extra tips

Keep Waxflower in water as much as possible as it drops flowers and leaves when out of water. Do not mist with water—flowers are very susceptible to fungal diseases such as *Botrytis*, which spread with misting.

GERBERA

Botanical name

Gerbera jamesonii.

Common name

Gerbera, Transvaal Daisy.

What to look for

Buy flowers that are fully open—Gerberas do not open further after they are picked. Most cultivars are harvested when the outer two rows of florets in the centre of the flower have begun to open and are showing pollen. The brown base of each stem should have been cut off. Flower centres should be perfectly round, not oval, and there should be no sign of petal damage (e.g. creases or dark marks) as Gerberas are prone to physical damage during transport. Gerberas for sale must be in clean water in clean buckets as they are very sensitive to unhygienic conditions.

Temperature

4 to 8°C.

Water

Recut 2 cm off each stem with sharp secateurs. Recut stems again if they are held out of water for more than 15 minutes, as air bubbles readily form in Gerbera stems, preventing water uptake. Change water daily to avoid bacterial build-up. Gerberas are very sensitive to bacterial plugging—always use very clean containers and fresh preservative.

Food

Preservative is essential.

Ethylene

Not sensitive.

Extra tips

Handle with extra care, as Gerbera petals are easily damaged. You can minimise damage by suspending flowers over the preservative solution—leave the flowers in the cardboard tray (if purchased that way) or put chicken-wire, or another support, over the top of a bucket and suspend flowers in the water. This may also help wilted Gerberas recover, as the heavy flower is supported and will not flop over, and the flowers will be able to take up more water. Some Gerbera varieties, namely older whites, yellows and pale pinks, are extremely sensitive to fluoride. Fluoride damage appears as brown marks on the tips of petals. If this is a problem, either use deionised water or rainwater, or ask your supplier for a newer variety, as these are far less susceptible to fluoride damage. Gerbera flowers naturally bend towards strong light when they are in the vase. To prevent this, wind a thin piece of wire around the outside of the stem. Don't force a wire up the middle of the stem as this will hinder water movement and cause early wilting. Gerbera stems can rot in floral foam. Some varieties have a problem where the stem just under the flower withers after 2 or 3 days and the flower-head flops over. This seems to be more of a problem in autumn and spring. It is not known why this occurs, and there appears to be little a florist can do to prevent it, as it affects flowers from most growers, and appears to be related to the weather.

GINGER

Botanical name

Zingiber species, *Alpinia purpurata*.

Common name

Ginger.

What to look for

There is a variety of forms of Ginger. Buy when 'cones' are glossy. The small flowers that poke out of the cones wilt quickly, but can be easily pulled off. Avoid buying soft, rubbery flowers, those with brown shrivelled tips, those that exhibit blueing or have signs of mould. Also avoid flowers or cones with black or brown marks.

Temperature

Above 12°C. *Do not refrigerate.*

Water

Recut 2 to 3 cm off each stem. Soak the whole stem and flower in warm water to revive wilted or shrivelled stems.

Food

Preservative is optional.

Ethylene

Not sensitive.

Ginger Lily

Botanical name

Hedychium gardnerianum.

Common name

Ginger Lily.

What to look for

Buy when the flowers on the spike are just starting to open. Fully open spikes do not last long. Check for scent as this is a sign of freshness.

Temperature

4 to 8°C.

Water

Recut 2 cm off stems and strip any leaves that would be underwater.

Food

Preservative is essential. It will help maintain open flowers and scent.

Ethylene

Not sensitive.

GLADIOLUS

Botanical name

Gladiolus hybrids.

Common name

Gladioli, Spear Lily, Sword Lily.

What to look for

Buy when the bottom two
florets in the spike are showing
strong colour, and can be
slightly open. At least 5 buds
up the spike should be showing
clear colour. The tips of the
spikes may be a little crooked. This is not a big problem as most
Gladdies are transported lying flat, and the tips continue to grow
upwards. Avoid spikes with brown marks on the sheaths covering
flowers and buds, or with yellow, dried leaves. Spikes with most of the
flowers open may be cheap, but they will not last long.

Temperature

4 to 8°C.

Water

Recut 2 to 4 cm off each stem with sharp secateurs. Gladdies are
transported dry, so it is important to cut a good 4 cm off each stem end,
then place directly in preservative solution. Strip any leaves that would
be underwater as leaves will rot quickly. Check water and replenish
regularly—Gladdies are heavy drinkers. It will also help to avoid
bacterial build-up.

Food

Preservative is essential. It will help buds open.

Ethylene

Mildly sensitive.

Extra tips

To straighten bent tips, place stems upright in water containing a preservative at 4°C for a few days. Some florists recommend removing the top two buds to reduce curvature and to encourage a more even opening of all the florets up the spike. Some *Gladiolus* varieties are also sensitive to fluoride in tap water, which results in tips of petals looking bleached. Other symptoms of fluoride toxicity are failure of the florets to open and develop normally.

GLORIOSA LILY

Botanical name

Gloriosa rothschildiana, Gloriosa superba.

Common name

Gloriosa, Rothschild Lily, Flame Lily, Glory Lily.

What to look for

Gloriosa are usually sold as single open flowers or as branches with one open flower and several buds. Buy open flowers that have strong red and yellow colouring, and no sign of brown spots. It doesn't matter if the anthers (containing pollen) remain or not. Branches should have one to two open flowers and several buds, which open well in the vase. Leaves should be glossy green, and should not have black marks, which would mean they have been refrigerated.

Temperature

Above 12°C. *Do not refrigerate.*

Water

Recut 2 cm off each branch with sharp secateurs. Strip any leaves that would be underwater as leaves will rot quickly. Change water regularly to avoid bacterial build-up.

Food

Preservative is essential.

Ethylene

Not sensitive.

Extra tips

Limp flowers should be revived by complete immersion in lukewarm water (about 30°C) for a few minutes. Be careful when separating stems as the ends of leaves have tendrils that often curl tightly around each other.

Godetia

Botanical name

Clarkia amoena.

Common name

Godetia, Satin Flower.

What to look for

Buy when flowers are in coloured
bud or half-open. Choose bunches
with many buds as these will open
in the vase. Avoid bunches with
tight green buds as they will not
open. Leaves should have no sign
of yellowing.

Temperature

1 to 4°C.

Water

Break open bunches, recut 2 cm off stems and strip any leaves that
would be underwater. Wash stems under the tap, then place in cold
water. Keep in water as much as possible, as Godetia dries out quickly.

Food

Preservative is essential. It will help maintain open flowers.

Ethylene

Sensitive.

GOMPHRENA

Botanical name

Gomphrena globosa.

Common name

Globe Amaranth, Strawberry Fields.

What to look for

Buy bunches when the red flowers are fully coloured. More mature flowers show flecks of pale pollen, but they will still last well. Leaves should be uniformly green with no sign of yellow.

Temperature

1 to 4°C.

Water

Break open bunches and cut 1 to 2 cm off each stem with sharp secateurs, strip any leaves that would be underwater and wash stems under the tap. Place immediately in cool water treated with a flower preservative.

Food

Preservative is essential. The preservative will help buds develop and ensure open flowers last.

Ethylene

Not sensitive.

Grevillea

Botanical names

Grevillea species, particularly *G. whiteana, G. sessilis, G. robusta* varieties.

Common names

Grevillea.

What to look for

Choose flowers where a third to a half of the lower flowers on the spike are open, or are in the 'looper' stage, where the stigmas are shaped in a loop and have not fully extended.

Temperature

1 to 4°C.

Water

Recut 2 cm off stems, and strip any leaves that would be underwater. Change vase water regularly.

Food

Preservative is essential.

Ethylene

Very sensitive.

Extra tips

Old flowers at the base of the spike tend to shed flower parts as the spike ages, possibly due to ethylene. Flowers are most likely not treated with anti-ethylene agents, so take great care to keep them away from ethylene sources while transporting and when in the shop.

GYMEA LILY

Botanical name

Doryanthes excelsa.

Common name

Gymea, Giant Lily.

What to look for

Buy when the red buds at the top of the flower-head have extended outwards and are starting to open. Buds will continue to open when stems are placed in water. Avoid stems with a tight flower-head, darkened leaves, or dry open flowers.

Temperature

1 to 4°C, if you can fit them in the fridge!

Water

Recut 4 to 5 cm off stems with sharp secateurs or a saw. Place in water as soon as stems have been recut. Keep in water as much as possible and the flower-head will come out.

Food

Preservative is recommended.

Ethylene

Not known to be sensitive.

Extra tips

These huge flowers are often up to 3 m in length. As you would expect, they are thirsty drinkers, so top up water regularly. Break off each flower as it dies, as this will encourage other flower buds to open.

GYPSOPHILA

Botanical name

Gypsophila paniculata.

Common names

Gyp, Baby's Breath, Million Stars.

What to look for

Buy when at least a third of the flowers are open. Buds will open in the vase if a preservative is used. Avoid bunches with brown flowers as they have a fungal infection, or they are old. Do not buy bunches that are not in water, as Gyp dries out easily. Check the foliage and the water that bunches are kept in. Avoid bunches that are not in clean water and clean buckets.

Temperature

1 to 4°C.

Water

Recut 2 cm off each branch with sharp secateurs. Strip any leaves that would be underwater as leaves will rot quickly. Change water daily to avoid bacterial build-up. Bunches should be treated by the grower with an anti-ethylene agent.

Food

Preservative is essential.

Ethylene

Sensitive.

Extra tips

Separate bunches carefully to prevent damaging the delicate flowers. Pinch off dead blooms as they produce ethylene gas, which will damage the rest of the flowers. Do not mist as Gypsophila is prone to fungal infection.

HELENIUM

Botanical name

Helenium autumnale varieties.

Common name

Helenium, Sneezeweed.

What to look for

Buy when flowers are half- to fully open. Check bunches for broken stems. Leaves should have no sign of yellowing.

Temperature

1 to 4°C.

Water

Break open bunches, recut 2 cm off stems and strip any leaves that would be underwater. Wash stems under the tap, then place in cold water.

Food

Preservative is essential. It will help maintain open flowers.

Ethylene

Not sensitive.

HELICONIA

Botanical name

A range of species, including *Heliconia caribae* (Claws), *H. rostrata* (Sexy Pink), *H. psittacorum* (Parrot Flower).

Common name

Heliconia, Crab Claw, Sexy Pink, Parrot Flower, Wild Plantain.

What to look for

Buy fully developed flowers, as they will not continue to develop after harvest. Avoid flowers with brown or black markings, as this is usually caused by chilling injury.

Temperature

Above 12°C. *Do not refrigerate.*

Water

Recut 2 to 4 cm off each stem with sharp secateurs and place in clean water.

Food

Preservative is optional. There is no evidence to suggest that preservative solutions extend flower life in Heliconias. However, it is recommended that chlorine is included in vase water to prevent bacterial growth.

Ethylene

Not sensitive.

Extra tips

Try and hold at high relative humidity (i.e. keep away from air-conditioners). Soaking in lukewarm water and misting may help.

HELLEBORUS

Botanical name

Helleborus niger, H. orientalis.

Common name

Christmas Rose, Lenten Rose.

What to look for

Buy when flowers are half- to fully open, with no sign of brown regions on the petals.

Temperature

1 to 4°C. Keep refrigerated at all times.

Water

Recut stems, removing the seared/scalded stem section. Strip leaves and place immediately in cold water.

Food

Preservative.

Ethylene

Not sensitive.

Extra tips

Will last better in a vase than in foam. If wilted, soak in cold water overnight. Likes to be misted.

HIPPEASTRUM

Botanical name

Hippeastrum hybrids.

Common name

Hippeastrum.

What to look for

Buy when buds show clear colour or are just starting to open. Avoid flowers with split and rolled stem ends.

Temperature

1 to 4°C.

Water

Recut 2 to 3 cm off stems and place immediately in cold water.

Food

Preservative is essential.

Ethylene

Not sensitive.

Extra tips

An overnight treatment in a solution of 3 heaped teaspoons of sugar and 30 ppm chlorine per litre of water can help mend split and curled stems, but may blow flowers open.

HYACINTH

Botanical name

Hyacinthus orientalis.

Common name

Hyacinth.

What to look for

Hyacinths are usually sold in bunches of three, with the bulb either cut off entirely or trimmed to look square. If you can find them, buy Hyacinths with the bulbs still attached as they last longer. The top florets should still be in bud, and the lower florets should be open evenly, which means one side should not be more open than the other. Strong fragrance is the best indicator of freshness. Avoid Hyacinths where all the florets are open, or those flowers that are leggy, with distinct gaps between each floret.

Temperature

1 to 4°C.

Water

Recut 2 cm off each stem with sharp secateurs and place in cold water.

Food

No preservative. Hyacinths don't like sugar. Place in cold water treated with chlorine.

Ethylene

Not sensitive.

Extra tips

Avoid placing Hyacinths near heat. They are winter-flowering bulbs and love the cold. Never mist flowers as this can cause fungal infections.

HYDRANGEA

Botanical name

Hydrangea macrophylla, H. paniculata grandiflora.

Common name

Hydrangea.

What to look for

Buy bunches with clean red, pink, blue or green colouring. Avoid bunches with brown flowers. Stems should have a clean cut at the end, and should not be seared, bashed or split.

Temperature

1 to 4°C.

Water

Recut stems and strip leaves that would be underwater. Place immediately in cold water.

Food

Preservative is essential.

Ethylene

Not sensitive.

Extra tips

Flower colour is determined by soil pH blue/mauve grow in acid soils, pink/red in alkaline soils. For drying, use flowers cut late in the season. Put them in 3 cm of water and do not add any extra water. Keep them out of sunlight to prevent fading.

Hymenocallis

Botanical name

Hymenocallis X *festalis*.

Common name

Spider Lily, Sacred Lily of the Incas.

What to look for

Buy when flowers are in bud but fully coloured (pale) or are half-open. Avoid bunches with yellow leaves.

Temperature

4 to 8°C.

Water

Recut 2 cm off stems and strip any leaves that would be underwater.

Food

Preservative is essential. It will help buds come out and maintain open flowers.

Ethylene

Not sensitive.

Hypericum

Botanical name

Hypericum species,
usually *H. chinense*.

Common name

Hypericum,
St John's Wort.

What to look for

Buy either when in
flower, or later when
berries are present. Flowers
need to be at least half-open,
and the berries shiny yellow and
red and plump. Foliage must be
glossy and showing no sign of
yellowing.

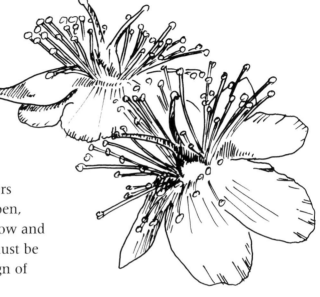

Temperature

1 to 4°C.

Water

Recut 2 cm off branches with sharp secateurs and strip any leaves that
would be underwater.

Food

Preservative is essential. It will help maintain open flowers.

Ethylene

Not sensitive.

Extra tips

Two interesting points on Hypericum it is a noxious weed in most
states, and certain forms are used as a herbal anti-depression treatment.

HYPOCALYMMA

Botanical name

Hypocalymma angustifolium, H. robustum.

Common name

Hypocalymma.

What to look for

Buy when the top third to a half of the flowers have opened. Flowers open from the tip of the branches down. Avoid bunches with closed flowers (not in bud) and with yellow leaves that are dropping.

Temperature

1 to 4°C.

Water

Recut 2 cm off branches with sharp secateurs and strip any leaves that would be underwater.

Food

Preservative is essential. It will help maintain open flowers.

Ethylene

Not sensitive.

Iris

Botanical name
  Iris hybrids.

Common name
  Dutch Iris.

What to look for
  Buy Dutch Iris when colour is visible
  from beneath the sheath covering the
  flower, but before petals have started to
  unfurl. As a general rule, it is better to
  buy Iris more open in mid-winter. For
  example, 'Professor Blaauw' (deep blue
  in colour) should be bought when the
  edge of one petal is clearly unfurled.
  Dried leaf tips should be trimmed. Avoid
  bunches with tips that are curled or dried.

Temperature
  1 to 4°C.

Water
  Recut 2 cm off each stem with sharp secateurs
  and place in clean water. Remove foliage that
  would be underwater. If flowers do not open, recut and wrap stems, but
  leaving heads unwrapped, and place in warm water (40°C) for no more
  than 30 minutes. Keep an eye on them as they may open very quickly.

Food
  No preservative. Iris don't like sugar. Place in cold water treated with
  chlorine.

Ethylene
  Mildly sensitive greyish-black petal tips indicate damage.

Extra tips
  Store upright as tips can curl upwards if placed flat. Keep away from
  draughts as flowers desiccate easily. Do not arrange with *Narcissus*
  (Daffodils) unless the Daffodils have been placed in a separate vase for
  a day.

ISOPOGON

Botanical name

A range of *Isopogon* species e.g. *I. cuneatis*.

Common name

Isopogon, Drumsticks.

What to look for

Buy bunches when the long, thin purple flowers at the base of the flower-head are starting to extend. The first whorl of flowers at the base of the cone should be starting to come out. Look for bent or broken flowers as they are easily damaged. Avoid bunches with brown tips on flowers, or where the thin flowers have wilted.

Temperature

1 to 4°C.

Water

Break open bunches and cut 1 to 2 cm off each stem with sharp secateurs, strip any leaves that would be underwater and wash stems under the tap. Place immediately in cool water treated with a flower preservative.

Food

Preservative is essential. The preservative will help buds to develop and ensure open flowers last.

Ethylene

Not sensitive.

IXODIA

Botanical name
Ixodia achilleoides ssp. *alata*.

Common name
South Australian Daisy, Hills Daisy.

What to look for
Avoid brown flowers. Flowers close when in low light but reopen readily once replaced in bright conditions, so don't worry if flowers are closed.

Temperature
1 to 4°C.

Water
Recut 2 cm off stems, strip leaves. Change vase water regularly.

Food
Preservative is recommended.

Ethylene
Not sensitive.

Extra tips
An excellent dried flower dries readily and retains colour and flower shape. When drying, leave out of water and hang upside down in a dark, cool and dry space.

Kangaroo Paw

Botanical name

Anigozanthos varieties or *Macropedia* species (Black Paw).

Common name

Kangaroo Paw.

What to look for

The first one or two florets in each spray should be open. The top buds should be plump i.e. not shrivelled. The first sign of ageing in Paws is when the top bud or flower flops over.

Temperature

1 to 4°C.

Water

Recut at least 2 cm off stems, strip all leaves off, if any, and place in cold water with a preservative.

Food

Preservative is essential. Kangaroo Paws love sugar use a commercial preservative or mix 3 or 4 teaspoons of sugar per litre of water along with 30 ppm chlorine.

Ethylene

Not sensitive.

Extra tips

Keep away from sunlight, draughts and heat sources as flowers fade and dry out easily.

KNIPHOFIA

Botanical name

Kniphofia uvaria.

Common name

Red Hot Poker.

What to look for

Buy when over half of the florets are open and coloured.

Temperature

1 to 4°C.

Water

Place in water treated with a preservative as soon as possible, after stripping any leaves that would be underwater and recutting 2 to 3 cm off each stem.

Food

Preservative is essential.

Ethylene

Very sensitive.

Extra tips

Store upright, as tips bend upwards if placed horizontally.

LACHNOSTACHYS

Botanical name

Lachnostachys verbascifolia.

Common name

Lambs' Tails.

What to look for

Buy bunches when flowers are in bud or open. Flowers are the tiny purple 'dots' running up the fluffy white/green flower-heads. Buy bunches with plump, full flower-heads, with no sign of wilting at the tip.

Temperature

1 to 4°C.

Water

Break open bunches and cut 1 to 2 cm off each stem with sharp secateurs, strip any leaves that would be underwater and wash stems under the tap. Place immediately in cool water treated with a flower preservative.

Food

Preservative is probably useful. The preservative will help buds develop and ensure open flowers last.

Ethylene

Not sensitive.

Lavender

Botanical name
Lavendula spica.

Common name
English Lavender, French Lavender.

What to look for
Buy when flowers are open and showing plenty of colour. Avoid dry flowers.

Temperature
4 to 8°C.

Water
Recut stems.

Food
Preservative.

Ethylene
Not sensitive.

LEPTOSPERMUM

Botanical name

Leptospermum species usually *L. scoparium* hybrids.

Common name

Tea Tree, Lavender Queen

What to look for

Buy when half of the flowers are open. Buds do not tend to open in the vase, so avoid bunches with closed buds. Flower drop becomes a problem with old bunches and late in the season. Shake branches to check for flower drop.

Temperature

1 to 4°C.

Water

Recut 2 cm off stems, strip any leaves that would be underwater.

Food

Preservative is essential.

Ethylene

Sensitive.

LEUCADENDRON

Botanical names

Leucadendron salignum, L. laureolum, L. discolor, L. salicifolium, L. macowanii, L. orientale, L. argenteum, L. comosum, L. coniferum.

Common names

Silvan Red, Safari Sunset, Inca Gold, Yellow Tulip, Flame Tip, Christmas Cones, Multi-cones.

What to look for

Buy bunches with straight stems and fresh leaves. Avoid stems with dried leaf tips.

Temperature

1 to 4°C.

Water

Recut 2 cm off stems, and strip any leaves that would be underwater.

Food

Preservative.

Ethylene

Not sensitive.

Extra tips

Can be stored for several days in clean water in the coolroom. Very thirsty flowers—keep well watered.

LEUCOSPERMUM

Botanical names

Leucospermum cordifolium, L. tottum, L. lineare, L. conocarpodendron.

Common names

Pincushion, Firewheel, Hawaii Gold, Sunburst, Catherine Wheel.

What to look for

Avoid crushed or bent flowers, and look for well-developed flower-heads. Leaf tips tend to turn brown with age.

Temperature

1 to 4°C.

Water

Recut 2 cm off stems, and strip any leaves that would be underwater.

Food

Preservative is recommended.

Ethylene

Not sensitive.

LIATRIS

Botanical name
 Liatris spicata.

Common name
 Liatris, Gayfeather, Blazing Star.

What to look for
 Buy when the entire stem is firm and the top buds have just opened. Liatris flowers open from the top downward, unlike most flowers which bloom from the bottom up. Do not expect all florets on the spike to open.

Temperature
 1 to 4°C.

Water
 Recut 2 cm off each stem with sharp secateurs and place in clean water. Remove foliage that would be underwater. Change the water daily Liatris pollutes water quickly.

Food
 Preservative. This will help flowers to come out.

Ethylene
 Not sensitive.

LILAC

Botanical name

Syringa vulgaris.

Common name

Lilac.

What to look for

A strong scent is the best indicator of freshness. Buy when most flowers (over half) are out. Shake branches, and avoid those where there is distinct flower drop. Avoid buying branches with marked and spotted flowers.

Temperature

1 to 4°C.

Water

Strip most of the leaves off the stem to minimise water loss, but leave the uppermost leaves on. Recut 2 cm off each stem with sharp secateurs—do not bash the stems. Top up water frequently.

Food

Preservative. This will help maintain scent.

Ethylene

Sensitive.

LILY

Botanical name

Lilium species *L. longiflorum* (Christmas Lily), *L. auratum* and *L. speciosum* hybrids (Orientals), *L. longiflorum* X Asiatic (LA Lilies), *L. regale* (Regals).

Common names

Christmas Lily, Oriental Lily, Asiatic Lily, LA Lily, Regal Lily.

What to look for

Buy Asiatics and Orientals when at least two buds per stem are fully developed and about to open, with the other buds well coloured. Christmas Lilies should be fully coloured white and in bud. Buds of all Lily types open well in the vase, but fully open flowers do not last nearly as long. Foliage should be glossy dark green and free of mottling or yellowing. Avoid the problem of split flowers in Christmas Lilies, which is especially common in December. Check cut stem ends—they should be green and crisp, not white and dry (indicating that the stems have been stored dry). If it is present, remove the lower white section of the stem, which does not take up water well.

Temperature

1 to 4°C.

Water

Recut at least 2 cm off each stem with sharp secateurs. Strip any leaves that would be underwater. Top up water frequently. Do not allow Lilies to dry out—place in water immediately.

Food

Preservative is essential. This will help open buds, maintain open flowers and retain scent in *L. longiflorum* and Orientals. Recut stems every few days and replace preservative.

Ethylene

Orientals and Asiatics are very sensitive. Other forms are mildly sensitive.

Extra tips

You can remove anthers (which contain the pollen) if desired, but this may reduce vase life. Pollen stains clothing and is best removed with sticky tape.

Lily-of-the-Valley

Botanical name

Convallaria majalis.

Common name

Lily-of-the-Valley.

What to look for

Buy when the first terminal bell has lost its deep green colour and is showing white.

Temperature

1 to 4°C.

Water

Place in water treated with a preservative as soon as possible, after stripping any leaves that would be underwater and recutting 1 cm off each stem, if possible. Place on ice if limp.

Food

Preservative is essential.

Ethylene

Not sensitive.

LISIANTHUS

Botanical name

Eustoma grandiflorum or Lisianthus russelianum.

Common name

Lisianthus, Prairie Gentian, Texas Blue-bell.

What to look for

Buy straight stems with healthy, glossy leaves and undamaged blooms. Each stem should have at least one open flower and many (at least 4) coloured buds. Avoid bunches with yellow foliage, creased flowers and those with wilted (hooked) tips.

Temperature

4 to 8°C.

Water

Recut 2 cm off each stem with sharp secateurs and place in clean water. Remove foliage that would be underwater.

Food

Preservative is essential. This will help the buds to come out.

Ethylene

Sensitive.

Extra tips

Place bunches in strong light, as this will help buds and flowers to develop full colour. If you do this, remember to check water regularly as they are thirsty drinkers.

Love-in-the-Mist

Botanical name

Nigella damascena varieties.

Common name

Love-in-the-Mist.

What to look for

Buy when flowers are fully coloured but before the petals have separated from the centre of the flower. Check bunches for broken stems. Leaves should have no sign of yellowing.

Temperature

1 to 4°C.

Water

Break open bunches, recut 2 cm off stems and strip any leaves that would be underwater. Wash stems under the tap, then place in cold water. Replace water daily.

Food

Preservative is essential. It will help maintain open flowers.

Ethylene

Not sensitive.

LUPIN

Botanical name

Lupinus varieties, mainly Russell Lupins.

Common name

Lupins.

What to look for

Buy when at least half of the flowers on the spike are open. Leaves should have no sign of yellowing.

Temperature

1 to 4°C.

Water

Recut 2 cm off stems and strip any leaves that would be underwater. Wash stems under the tap, then place in cold water.

Food

Preservative is essential. It will help maintain open flowers.

Ethylene

Mildly sensitive.

LYSIMACHIA

Botanical name

Lysimachia ephemerum.

Common name

Loosestrife.

What to look for

Buy bunches when about a third to a half of the lower flowers in the spike are open. Look for straight stems, as tips tend to wilt. Avoid bunches with yellow or black leaves at the stem ends.

Temperature

1 to 4°C.

Water

Break open bunches and cut 1 to 2 cm off each stem with sharp secateurs, strip any leaves that would be underwater and wash stems under the tap. Place immediately in cool water treated with a flower preservative.

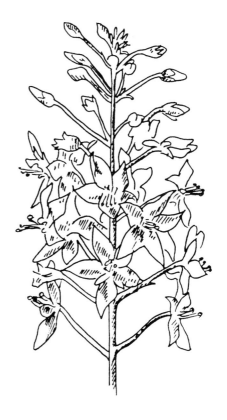

Food

Preservative is essential. The preservative will help buds to develop and ensure open flowers last.

Ethylene

May be sensitive.

MARIGOLD

Botanical name

Tagetes patula varieties.

Common name

French Marigold.

What to look for

Buy bunches with fully open flowers and fresh green leaves.

Temperature

1 to 4°C.

Water

Break bunches apart, recut 2 cm off stems and strip any leaves that would be underwater. Wash stripped stems under the tap, and place in cold water.

Food

Preservative is essential. It will help maintain open flowers.

Ethylene

Not sensitive.

MATRICARIA

Botanical name

Matricaria eximia.

Common name

Feverfew.

What to look for

Buy bunches when petals are open and the outer half of the yellow flower centre is rough i.e. when about half of the small flowers in the yellow centre are open. Avoid bunches with yellow or black leaves at the stem ends.

Temperature

1 to 4°C.

Water

Break open bunches and cut 1 to 2 cm off each stem with sharp secateurs. Strip any leaves that would be underwater and wash stems under the tap. Place immediately in cool water treated with a flower preservative.

Food

Preservative is essential. The preservative will help buds develop and ensure open flowers last.

Ethylene

Not sensitive.

Misty

Botanical name

Limonium latifolium X *bellidifolium*.

Common name

Misty Blue, Misty Pink.

What to look for

Buy when at least half of the tiny flowers are open, as buds do not open in the vase. Avoid bunches with dark foliage, or those with a distinct rotten-egg smell.

Temperature

1 to 4°C.

Water

Place in cold water treated with a preservative as soon as possible, after stripping any leaves that would be underwater and recutting 2 to 3 cm off each stem. Replace with fresh cold water daily to minimise the smell.

Food

Preservative is essential as it helps prevent the rotten-egg smell from developing.

Ethylene

Not sensitive.

Molucella

Botanical name
Molucella laevis.

Common name
Bells of Ireland.

What to look for
Buy when the bells are open and are light green in colour. Flowers should not be losing the white flecks of pollen. Avoid bunches with yellowing bells.

Temperature
1 to 4°C.

Water
Place in water treated with a preservative as soon as possible, after stripping any leaves that would be underwater and recutting 2 to 3 cm off each stem, if possible.

Food
Preservative is essential.

Ethylene
Not sensitive.

MUSCARI

Botanical name

Muscari botryoides.

Common name

Grape Hyacinth.

What to look for

Buy when flowers are open and clear blue in colour.

Temperature

1 to 4°C.

Water

Muscari bunches are very short, but trim 1 cm off stems if you can. Wash stems under the tap, then place in cold water.

Food

Preservative is essential. It will help maintain open flowers.

Ethylene

Not sensitive.

Narcissus

Botanical name

Narcissus hybrids, *N. tazetta* (Paperwhite).

Common names

Daffodil; Jonquil Paperwhite, Erlicheer.

What to look for

Buy Daffodils when the flowers are in bud and showing clear colour but petals have not bent back from the trumpet (gooseneck stage). Flowers bought fully open will not last as long. Jonquils are best bought at the 'one bell' stage, when one flower is open in each cluster. Later on in the season, however, it is difficult to buy anything other than open flowers, which should have a 'crisp' feel and the trumpet should be free from brown marks. Check the cut stem ends and avoid bunches where these have split and curled upwards.

Temperature

1 to 4°C. Keep cool at all times.

Water

Recut 2 cm off each stem with sharp secateurs and place in clean water by themselves. Never combine fresh Daffodils and Jonquils with other flowers, especially Roses, Carnations, Freesias, Tulips and Anemones. All *Narcissus* varieties exude a toxic sap when the stems are freshly cut. The sap is harmful to other cut flowers. After recutting, put *Narcissus* alone in a container filled with cold water for at least 24 hours before arranging with other flowers. Do not recut stems. Discard water and wash the container thoroughly.

Food

No preservative. Place flowers in cold water treated with chlorine only.

Ethylene

Not sensitive.

Extra tips

Always store upright as the stems will bend upright if laid flat.

NERINE

Botanical name
Nerine bowdenii, Nerine sarniensis.

Common name
Nerine, Guernsey Lily, Cape
Colony Nerine.

What to look for
Buy when the oldest bud in
each stem is partially to fully
open. Avoid very tight buds as
they can develop crooked stems.

Temperature
6 to 10°C.

Water
Nerines are prone to drying out, so put them in water promptly. Recut
2 cm off each stem with sharp secateurs and place in clean water.
Remove any foliage that would be underwater.

Food
Preservative is essential. This will help buds open.

Ethylene
Mildly sensitive.

ORCHID

Botanical names

Mostly *Cymbidium, Dendrobium,*
(Singapore Orchid), *Oncidium* and
Vanda, rarely *Phalaenopsis* (Moth
Orchid) or *Cattleya.*

Common names

Cymbids, Singapore Orchid,
Moth Orchid.

What to look for

Fully open to half-open flowers,
as buds will not develop after
sprays are cut. Try to buy sprays
with stem ends placed in vials full of preservative solution or damp
cotton wool. With the exception of some Cymbidiums, Orchids are
imported, so check flowers carefully to ensure they are not creased or
bruised. Avoid flowers with a slightly dried, transparent appearance, as
this indicates chilling or ethylene injury.

Temperature

Above 12°C. *Do not refrigerate.*

Water

Recut at least 2 to 4 cm off stems if they are bought without stem ends
in vials, or if the water in the vials has run out. Either top up the
solution in the vials and replace cut stems, or place the stems in vases
with clean water containing preservative. Refill vial solutions daily.

Food

Preservative is essential. This will help flowers open and keep open
flowers looking fresh.

Ethylene

Very sensitive.

Extra tips

Individual Dendrobium flowers can be immersed in room-temperature tap water for 5 minutes to revive them. Individual Orchid flowers can then be placed in a plastic bag with a piece of moist paper and kept above 12°C for a few days. Handle very carefully as flowers are susceptible to bruising. Rough handling of Cymbidium flowers can also dislodge the pollen cap, which will result in rapid wilting. Misting is recommended.

ORNITHOGALUM

Botanical name

Ornithogalum arabicum (Star of Bethlehem), *O. thyrsoides* (Chincherinch).

Common name

Star of Bethlehem, Chincherinch.

What to look for

Buy bunches when about a quarter of the lower flowers on the spike are open. Avoid bunches with wilted lower flowers and yellow leaves.

Temperature

1 to 4°C.

Water

Break open bunches and cut 1 to 2 cm off each stem with sharp secateurs, strip any leaves that would be underwater and wash stems under the tap. Place immediately in cool water treated with a flower preservative.

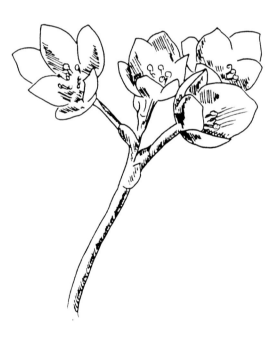

Food

Preservative is optional.

Ethylene

Not sensitive.

PEONY

Botanical name
Paeonia lactiflora hybrids.

Common name
Peony, Tree Peony.

What to look for
Buy flowers when the first petal is separating from the bud and colour is clearly visible. Flowers cut earlier when buds are completely green will have a shorter vase life. Fully open flowers will last about half as long as coloured buds and are very prone to physical damage, so watch out for bent, creased or torn petals.

Temperature
6 to 10°C.

Water
Recut 2 cm off each stem with sharp secateurs and place in clean water. Remove any foliage that would be underwater.

Food
Preservative is essential. This will help buds open.

Ethylene
Not sensitive.

PHLOX

Botanical name

Phlox paniculata, P. drummondii.

Common name

Phlox, Wild Sweet William.

What to look for

Buy when over half of the florets are open. Avoid bunches with signs of flower drop.

Temperature

1 to 4°C.

Water

Recut stems, and strip any leaves that would be underwater. Place immediately in cold water.

Food

Preservative is essential. It will help buds develop and help delay flower drop.

Ethylene

Very sensitive.

PHYSOSTEGIA

Botanical name

Physostegia purpurea, P. virginiana.

Common name

Physostegia.

What to look for

Buy when at least 3 flowers per spike are open. Only buy bunches that have been treated against ethylene as Physostegia is extremely sensitive. Avoid bunches with flower drop. Pink varieties seem to last longer.

Temperature

1 to 4°C.

Water

Place in water treated with a preservative as soon as possible, after stripping any leaves that would be underwater and recutting 2 to 3 cm off each stem. Replace with fresh water daily.

Food

Preservative is essential.

Ethylene

Very sensitive.

Pieris

Botanical name

Pieris japonica.

Common name

Andromeda, Pieris.

What to look for

Buy bunches when the bell-shaped flowers are fully open. Leaves should be glossy green, and flowers unmarked.

Temperature

1 to 4°C.

Water

Recut 2 cm off branches with sharp secateurs and strip any leaves that would be underwater.

Food

Preservative is essential. It will help maintain open flowers.

Ethylene

Not sensitive.

Pineapple Lily

Botanical name

Eucomis comosa.

Common name

Pineapple Lily.

What to look for

Buy when the flowers in the lower third to half of the spike are open. Look for straight stems.

Temperature

4 to 8°C.

Water

Recut 2 cm off stems and strip any leaves that would be underwater.

Food

Preservative is essential. It will help buds to open up the spike, and maintain open flowers.

Ethylene

Mildly sensitive.

PROTEA

Botanical names

A wide range of Proteas are used as cut flowers, including *Protea neriifolia*, *P. cynaroides*, *P. magnifica*, *P. grandiceps*, *P. repens*, *P. eximia*, *P. scolymocephala*.

Common names

Pink Mink, White Mink, Pink Ice, King Protea, Queen Protea, Princess Protea, Honey Protea, Mini Protea.

What to look for

Buy Proteas with bracts that are starting to open *P. neriifolia* and Pink Ice should have a circle at least 4 cm wide at the top, while the bracts in Kings should be reflexed back. Avoid flowers with a grey tinge as they are old. Most importantly, check the leaves for signs of blackening in Kings, Queens, *P. neriifolia* and Pink Ice—they should be uniformly green with no black or brown spots.

Temperature

1 to 4°C.

Water

Recut 2 cm off stems with sharp secateurs, and strip any leaves that would be underwater and place in cold water. Change the vase water at least every second day, as Proteas turn vase water brown.

Food

Preservative is recommended, as it helps delay leaf blackening.

Ethylene

Not sensitive.

Extra tips

Keep Proteas under strong light, as this reduces leaf blackening.

Ptilotus

Botanical name
Ptilotus obovatus.

Common name
Mulla Mulla.

What to look for
The feathery flower-heads should have started to open. Look for strong stems that can hold the flower-heads. Flower-heads should be showing pink-white in colour.

Temperature
1 to 4°C.

Water
Recut 2 to 3 cm off each stem. Place in cold water immediately.

Food
Preservative is essential.

Ethylene
Not sensitive.

QUEEN ANNE'S LACE

Botanical name

Ammi majus.

Common name

Queen Anne's Lace.

What to look for

Buy when the white flowers are half- to fully open. Stems must be thick enough to support the flower-heads. Avoid bunches with wilted heads and yellow leaves.

Temperature

1 to 4°C.

Water

Place in cold water treated with a preservative as soon as possible, after stripping any leaves that would be underwater and recutting 2 to 3 cm off each stem.

Food

Preservative is essential.

Ethylene

Not sensitive.

Extra tips

Tends to wilt if long stems are used in floral foam.

RANUNCULUS

Botanical name

Ranunculus asiaticus.

Common name

Persian or Turban Buttercup.

What to look for

Buy when the flowers are in bud and fully coloured. Avoid bunches with yellow, decayed leaves.

Temperature

1 to 4°C.

Water

Split bunches and thoroughly wash stems. Recut stems and strip leaves as they will foul the water quickly. Change vase water regularly.

Food

Preservative.

Ethylene

Not sensitive.

RICEFLOWER

Botanical name

Ozothamnus diosmifolius, Cassinia species.

Common name

Riceflower.

What to look for

Buy fresh branches, with tips that are not dry and curled. Shake bunches to test for signs of leaf and flower drop. Leaves tend to dry and drop before the 'flowers'.

Temperature

1 to 4°C.

Water

Recut 2 cm off stems and strip any leaves that would be underwater. Place in cold water.

Food

Preservative is essential. It will help delay leaf and flower drop.

Ethylene

Not sensitive.

Extra tips

First sign of ageing is leaf yellowing and drop.

ROSE

Botanical name

Rosa hybrida.

Common name

Rose.

What to look for

Buy bunches with strong stems that
support the blooms. It is impossible to
suggest a perfect opening stage to buy
Roses, as this varies quite widely between
varieties. In general, vase life will be better for
flowers picked a little more open than those sold in tight bud—this is
especially true in winter. Tight buds can fail to open, particularly during
the colder months, and are prone to bent neck, where the stem
weakens just under the flower and the flower-head flops over. Lower
leaves should be dark green and the veins should not be prominent.
Check the cut stem ends—they should be crisp and green, not dry and
dark, which is a sure sign of age. Roses suffer from *Botrytis*, a fungal
infection that turns petals brown; extra care is needed when the
weather is warm and humid. While it is often difficult to detect *Botrytis*,
look closely at the outer petals and do not buy those bunches with
small brown spots. Once a Rose has this problem there is nothing you
can do other than throw the Rose away.

Temperature

1 to 4°C. Keep Roses in the fridge or the coolroom as much as possible,
with plastic placed over buckets—do not place outside the shop.

Water

Flower life in Roses is highly dependent on flowers being looked after
properly at every stage. If possible, recut stems underwater—removing
2 to 3 cm from each stem. If you can't cut underwater, cut with sharp
secateurs and immediately place stems in cold water treated with a
preservative. Top up water at least daily. In general, use cool water to
prevent early opening and 'blown' blooms, and use warm water to

hasten bud opening, but be careful as Roses can blow open within minutes in warm water. Strip any leaves that would be underwater. Only remove thorns when it is absolutely necessary. Remove thorns and leaves carefully so as not to damage stems, as this will increase the likelihood of bacterial infection.

Food

Preservative is essential. This will help buds open and maintain open flowers. Replace with fresh preservative at least every second day.

Ethylene

Roses are sensitive to ethylene, with some varieties, notably dark reds, being more sensitive than others. Signs of ethylene damage are blackened edges of petals and failure of buds to open.

Extra tips

A good way to revive wilted Roses is to recut stems (preferably underwater), then wrap the flower-heads tightly in newspaper, forcing the heads into an upright position. Place the stems, up to just below the flower-heads, in lukewarm water for several hours. This way the flowers are not underwater, which minimises the chance of fungal diseases spreading.

Rudbeckia

Botanical name

Rudbeckia hirta hybrids.

Common name

Rudbeckia, Black-eyed Susan.

What to look for

Buy bunches when flowers are fully open. Avoid bunches with yellow or black leaves at the stem ends.

Temperature

1 to 4°C.

Water

Break open bunches and cut 1 to 2 cm off each stem with sharp secateurs, strip any leaves that would be underwater and wash stems under the tap. Place immediately in cool water treated with a flower preservative.

Food

Preservative is essential. The preservative will help buds develop and ensure open flowers last.

Ethylene

Sensitive.

SANDERSONIA

Botanical name

Sandersonia auriculata.

Common name

Golden Bells.

What to look for

Choose bunches with between a third and a half of the lower flowers (bells) open. Avoid bunches with yellow leaves. If possible, buy bunches with the stem ends placed in sachets filled with water.

Temperature

4 to 8°C.

Water

Recut 2 cm off each stem with sharp secateurs and place in clean water. Remove any foliage that would be underwater. Alternatively, recut stems and replace in sachets filled with preservative.

Food

Preservative is essential. This will help buds open.

Ethylene

Not sensitive.

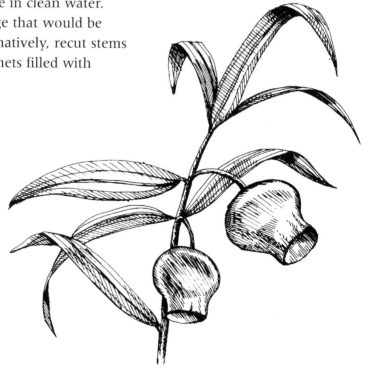

SCABIOSA

Botanical name

Scabiosa atropurpurea.

Common name

Scabiosa, Pincushion Flower.

What to look for

Buy when flowers are fully open. Try and buy bunches that have been treated against ethylene, although this could be difficult. Avoid bunches with petal drop and/or yellow leaves.

Temperature

1 to 4°C.

Water

Place in water treated with a preservative as soon as possible, after stripping any leaves that would be underwater and recutting 2 to 3 cm off each stem.

Food

Preservative is essential.

Ethylene

Very sensitive.

SEDUM

Botanical name

Sedum species, usually *S. spectabile*.

Common name

Sedum.

What to look for

Buy when flowers are forming a dense coloured display on the top of the stems and are all open. Leaves should show no sign of yellowing.

Temperature

1 to 4°C.

Water

Break open bunches, recut 2 cm off stems and strip any leaves that would be underwater. Wash stems under the tap, then place in cold water.

Food

Preservative is essential. It will help maintain open flowers.

Ethylene

Not sensitive.

SNAPDRAGON

Botanical name
 Antirrhinum majus.

Common name
 Snapdragon.

What to look for
 Buy when at least a third of the lower florets are open. Shake bunches
 to make sure all florets are intact and there is no drop. Snapdragons
 bend upwards when lain flat, so choose straight spikes. Check that your
 supplier has treated bunches against ethylene.

Temperature
 1 to 4°C.

Water
 Recut 2 cm off each stem with sharp secateurs and place in clean water.
 Remove any foliage that would be underwater as it rots quickly.
 Replace water daily.

Food
 Preservative is essential. This will help buds open.

Ethylene
 Very sensitive.

Extra tips
 Tips will bend upwards if stored flat, so always keep upright in water.
 Pinching the top bud when arranging flowers will prevent this bending
 and promote even flowering of the rest of the buds.

SOLIDASTER

Botanical names

Aster ericoides X *Solidaster luteus* (cross between Solidago and Aster).

Common names

Solidaster.

What to look for

Buy when at least half of the small yellow flowers are open. Avoid bunches with yellow leaves.

Temperature

1 to 4°C.

Water

Recut 2 cm off stems, strip leaves and place in cold water. Change vase water regularly.

Food

Preservative is essential.

Ethylene

Not sensitive.

STATICE

Botanical name

Limonium sinuatum,
L. suworowii.

Common name

Statice, Russian
Statice, Rat's Tails.

What to look for

Buy stems where most
of the flowers are fully
developed as they do not
open in the vase. Avoid
bunches with slimy stems and
lower leaves. Check the section of the bunch that is underwater, and
avoid those with a strong sulphur smell.

Temperature

1 to 4°C.

Water

Break apart bunches as soon as possible for better air circulation. Recut
2 cm off each stem, and strip leaves that will be underwater, then wash
stems under the tap to remove any dirt and slime. Change vase water
daily, as Statice quickly fouls it.

Food

Preservative is optional. However, chlorine in the vase water is essential
with Statice to control growth of bacteria and slime.

Ethylene

Not sensitive.

Extra tips

Statice can be dried easily by hanging bunches upside down in a well-
ventilated area.

STEPHANOTIS

Botanical name

Stephanotis floribunda.

Common name

Stephanotis.

What to look for

Buy when one or two flowers per cluster are open. Flowers should be pure white, with no sign of brown marks or creases. Check for strong fragrance as this is a sign of freshness.

Temperature

1 to 4°C.

Water

Recut 2 cm off stems and strip any leaves that would be underwater. Place in cold water.

Food

Preservative is essential. It will help buds come out and maintain scent.

Ethylene

Sensitive.

Extra tips

May be misted.

STIRLINGIA

Botanical name

Stirlingia simplex.

Common name

Stirlingia.

What to look for

When buying Stirlingia it is important to work out if the branches have been 'dried and dyed'. Dried branches do not need water, as they have been preserved and will last for months. Fresh Stirlingia should be bought with at least half the flowers open, as they don't open much in the vase.

Temperature

1 to 4°C.

Water

Keep dried and dyed branches out of water. For fresh branches, carefully strip any leaves and small branches that would be underwater, and cut at least 2 to 3 cm off the end of each branch. Place in cold water that contains a preservative. Keep cool as much as possible.

Food

Preservative is essential for fresh branches.

Ethylene

Not sensitive.

STOCK

Botanical name

Matthiola incana hybrids.

Common name

Stock.

What to look for

Buy stems with at least a third to a half of the lower flowers open. Avoid bunches with brown marks on petals, as Stock are susceptible to fungal diseases. Bunches should have had the woody, white root section cut off each stem. Check for scent—strong scent is a good sign of freshness.

Temperature

1 to 4°C.

Water

Recut 2 cm off stems with secateurs and be sure to remove the woody base and roots, then wash stems under the tap to remove any dirt and slime. Remove all leaves that would be underwater. Vase water containing Stock quickly develops a foul odour due to rapid bacterial growth. To control this, you must use a preservative and change water daily.

Food

Preservative is essential.

Ethylene

Sensitive.

Sunflower

Botanical name

Helianthus annuus.

Common name

Sunflower.

What to look for

Buy when flowers are half- to fully open. Look for strong stems that will support the (sometimes) large flower. Leaves should have no sign of wilting, black marks or yellowing.

Temperature

4 to 8°C.

Water

Recut 2 to 4 cm off stems and strip any leaves that would be underwater. Leaves will wilt and die before the flower, so it is a good idea to strip all but the top leaves. To help prevent leaf drooping, add 10 drops of household detergent to 5 L of water, and leave flowers in this solution for 1 to 3 hours, or no longer than overnight.

Food

Preservative is essential. It will help maintain open flowers.

Ethylene

Not sensitive.

SWEET PEA

Botanical name

Lathyrus odoratus hybrids.

Common name

Sweet Pea.

What to look for

Buy bunches where the top buds are fully coloured and partly open. Check for scent—strong scent is a good sign of freshness. Shake bunches and avoid those where flowers drop off, as this is a sign of ethylene damage. Avoid bunches with yellow leaves.

Temperature

1 to 4°C.

Water

Recut 2 cm off stems. Remove all leaves that will be underwater and place in cold water.

Food

Preservative is essential.

Ethylene

Very sensitive.

Extra tips

Take care when separating bunches, as Sweet Peas cling together with tendrils. Rough handling will damage flowers and leaves.

THRYPTOMENE

Botanical name

Thryptomene calycina.

Common names

Grampians Thryptomene, Australian Lace Flower.

What to look for

Avoid bunches with closed flowers and dry foliage. Choose bunches where a fifth to a third of flowers are open, and watch out for flower drop late in the season (September–October). Don't buy Thryptomene bunches that have been placed on the floor under wet hessian—this reduces vase life and can result in fungal infection.

Temperature

1 to 4°C.

Water

Separate bunches, recut 2 cm off stems with sharp secateurs and strip any leaves that would be underwater. Wash stripped stems under the tap, then place in cold water.

Food

Preservative is essential to open buds and maintain open flowers.

Ethylene

Not sensitive.

Extra tips

Will take up water through the leaves and flowers, so can be misted.

TRACHELIUM

Botanical name

Trachelium caeruleum.

Common name

Trachelium.

What to look for

Buy when a quarter to a third of the flowers are open. Only buy bunches that have been treated against ethylene. Avoid bunches with flower drop and/or yellow leaves.

Temperature

1 to 4°C.

Water

Place in water treated with a preservative as soon as possible, after stripping any leaves that would be underwater and recutting 2 to 3 cm off each stem.

Food

Preservative is essential.

Ethylene

Very sensitive.

Triteleia

Botanical name

Triteleia laxa.

Common name

Queen Fabiola, Brodiaea.

What to look for

Buy bunches when more than 6 flowers are open per stem and buds are showing a clear purple-blue colour. Avoid bunches with wilted lower flowers and yellow leaves.

Temperature

1 to 4°C.

Water

Break open bunches and cut 1 to 2 cm off each stem with sharp secateurs, making sure you remove the white stem base. Strip any leaves that would be underwater and wash stems under the tap. Place immediately in cool water treated with a flower preservative.

Food

Preservative is essential. The preservative will help buds develop and ensure open flowers last.

Ethylene

Very sensitive.

TUBEROSE

Botanical name

Polianthes tuberosa.

Common name

Tuberose.

What to look for

Buy bunches when 2 to 4 of the lower florets per stem are open. Check for scent—strong scent is a good sign of freshness. Avoid stems with wilted lower florets, browning of petal margins or floppy tips or stems—these are signs of ageing.

Temperature

4 to 8°C.

Water

Recut 2 cm off stems. Remove all leaves that will be underwater and place in warm water for up to 2 hours to open buds.

Food

Preservative is essential. Failure of buds to open can be a problem in Tuberose and preservative will help minimise this problem.

Ethylene

Very sensitive.

Extra tips

Some florists recommend removal of the top bud to reduce curvature and promote even opening of flowers. Remove wilted lower florets as they produce ethylene, which will affect the rest of the bunch.

Tulip

Botanical name

Tulipa hybrids.

Common name

Tulip, Parrot Tulip, Darwin Tulip.

What to look for

Buy bunches where flowers are not fully open, but the buds are fully developed and well-coloured (half colour, half green). This is dependent on variety— some white and pale yellow varieties can be bought when buds are still green and will develop fully. Stems must be strong enough to hold the flower-head upright. Foliage must be glossy and green, not blotchy (fungal disease), striped (virus), or yellowing (old). Tulips are quite tolerant of a little mud in the water.

Temperature

1 to 4°C. Keep cool at all times.

Water

Strip any leaves that would be underwater, then wash mud off the lower stems under the tap. Recut 2 cm off each stem and place in cold water containing chlorine. Be sure to remove as much of the white portion of the stem as is practical to improve water uptake.

Food

No preservative. The sugar in preservatives results in stem stretching, causing flowers to flop over. Use cold water with 30 ppm chlorine.

Ethylene

Not sensitive.

Extra tips

Tulip stems continue to grow towards the light, which can result in twisted stems. To straighten bent stems, wrap bunches firmly in dampened tissue paper then newspaper, place in clean cold water and leave in a cool spot for a few hours. Do not mix with freshly cut *Narcissus*.

UMBRELLA FERN

Botanical name

Sticheris flabellatus.

Common name

Umbrella Fern.

What to look for

Buy fully open fronds. Check for tears and creases on fronds and avoid damaged branches. Fronds should be fresh and have an even green colour with no sign of wilting.

Temperature

1 to 4°C.

Water

Recut 2 to 3 cm off each stem and place in cold water treated with chlorine.

Food

Preservative is not needed.

Ethylene

Not sensitive.

Vallota Lily

Botanical name
 Vallota speciosa.

Common name
 Vallota Lily, George Lily.

What to look for
 Buy when flowers are in bud, but fully coloured (red), or when half-open.

Temperature
 4 to 8°C.

Water
 Recut 2 cm off stems and strip any leaves that would be underwater.

Food
 Preservative is essential. It will help buds come out and maintain open flowers.

Ethylene
 Not sensitive.

VERONICA

Botanical name

Veronica virginica, V. spicata.

Common name

Veronica, Speedwell.

What to look for

Buy when about a third to a half of the flowers on the spike are open. Bunches with fully open spikes will not last as long as those with half-open spikes. Leaves should have no sign of yellowing.

Temperature

1 to 4°C.

Water

Break open bunches, recut 2 cm off stems and strip any leaves that would be underwater. Wash stems under the tap, then place in cold water.

Food

Preservative is essential. It will help maintain open flowers.

Ethylene

Very sensitive.

Verticordia

Botanical name

Verticordia species *V. nitens* (yellow), *V. grandis* (red), *V. grandiflora*
(white).

Common name

Feather Flower, Yellow Morrison (*V. nitens*).

What to look for

Choose bunches with fresh leaves as leaves often dry and drop before
the flowers. *Verticordia* is very susceptible to fungal infection—do not
buy bunches with the characteristic 'spider-web' caused by *Botrytis*.

Temperature

1 to 4°C.

Water

Separate bunches,
recut 2 cm off
stems with sharp
secateurs and strip
any leaves that
would be
underwater.

Food

Preservative is
recommended.

Ethylene

Very sensitive.

VIOLET

Botanical name

Viola odorata.

Common name

Violet.

What to look for

Buy when flowers are open with flat petals. Leaves should be fresh and green; avoid bunches whose leaves are yellow or marked with blotches. Check for scent as this is a sign of freshness.

Temperature

1 to 4°C.

Water

Violet bunches are very short, so trim only 1 cm off stems and strip any leaves that would be underwater. Place in cold water immediately.

Food

Preservative is essential. It will help maintain open flowers.

Ethylene

Not sensitive.

Extra tips

Misting or dunking violet bunches underwater is recommended, provided the water is clean. This is best done by adding a small amount of chlorine (about 10 ppm), as this will prevent transmission of fungal diseases.

WALLFLOWER

Botanical name

Cheiranthus cheiri.

Common name

Wallflower.

What to look for

Buy when flowers are fully open. Check bunches for broken stems. Leaves should show no sign of yellowing.

Temperature

1 to 4°C.

Water

Break open bunches and recut 2 cm off stems and strip any leaves that would be underwater. Wash stems under the tap, then place in cold water.

Food

Preservative is essential. It will help maintain open flowers.

Ethylene

Not sensitive.

WARATAH

Botanical name

Telopea speciosissima.

Common name

Waratah.

What to look for

Waratahs are composite flowers comprising up to 300 individual florets surrounded by coloured bracts. Early in the season, choose flowers with a few (up to 5%) of the florets open. Later on, choose flowers with up to a half of the florets open. A rule of thumb is to choose flowers with the least number of flowers open, as long as some are open. Flowers with over half of the florets open won't last as long. Avoid flowers with a blue tinge, as this is a sign of ageing and ethylene damage.

Temperature

1 to 4°C.

Water

Recut 2 cm off stems with sharp secateurs, and strip any leaves that would be underwater.

Food

Don't use a commercial preservative. Place in cold water with 30 ppm chlorine. The sugar contained in preservatives causes excessive nectar production which can lead to problems with wasps and fungal diseases.

Ethylene

Very sensitive.

Extra tips

Take precautions against ethylene Waratahs are very sensitive to ethylene but STS treatment does not appear to protect them.

ZANTEDESCHIA

Botanical name

Zantedeschia species *Z. elliottiana* (Golden Calla), *Z. rehmannii* (Pink Calla), *Z. aethiopica* (Arum).

Common name

Calla Lily, Arum Lily, Green Goddess.

What to look for

Buy flowers that are completely open, when the large fused petal (spathe) has unrolled and just before it has begun to turn downwards. Look for unmarked blooms—Callas are prone to creases and cuts as opened flowers are easily damaged. Avoid stems where the ends are split and curling upwards.

Temperature

4 to 8°C.

Water

Recut 2 cm off stems.

Food

Preservative is optional. Use cold water with 30 ppm chlorine as an alternative.

Ethylene

Not sensitive.

Extra tips

To straighten stems, suspend flowers from the neck with chicken wire and allow to stand in cold water overnight.

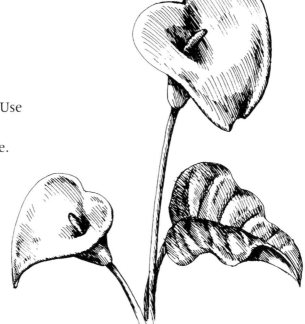

ZINNIA

Botanical name
  Zinnia elegans.

Common name
  Zinnia, Youth and Age.

What to look for
  Buy bunches that have fully open flowers with tight centres. Avoid flowers where the centres are full of yellow pollen as they are old.

Temperature
  1 to 4°C.

Water
  Recut 2 cm off stems. Strip leaves that will be underwater. Replenish water daily.

Food
  Preservative.

Ethylene
  Not sensitive.

GLOSSARY

abscise: To fall off. This term is normally used to describe premature loss of leaves or flowers.

acidifier: A substance that lowers the pH below 7.0 or makes water more acid. Acidifiers help prevent plugging and reduce microbial growth. Examples are vinegar, citric acid and alum (aluminium sulphate). Ideally, vase water should have a pH of between 3.0 and 4.0. Acidifiers are particularly beneficial for woody-stemmed flowers, including Roses.

air emboli: Bubbles of air that are drawn up the base of the flower stem after cutting, which effectively slow water uptake. They are usually located in the bottom 2 to 3 cm of the stem and do not move further up.

alkaline: Opposite to acid, where the pH is above 7.0. Some tap water supplies are very alkaline (known as hard water), which may mean that some floral preservatives do not work as expected.

alum: Aluminium sulphate. This is especially useful for Roses as an acidifier at a concentration of 50 ppm (0.05–0.01 g per litre or 1–2 teaspoons per 100 L). Alum acts by forming flocculent aluminium hydroxide, which traps particles in water and also lowers the pH.

ambient temperature: Room or surrounding air temperature.

bent neck: Rose flowers flop over just under the flower-head. This often happens with Roses during summer. There are usually three reasons for this
- the Rose is picked too early, before the top of the stem has developed enough woody tissue (lignin) to support the bloom;
- an embolism has prevented water uptake;
- microbes have plugged the stem.

Bent neck can be overcome by
- purchasing Roses that are slightly more open;
- recutting underwater;
- using an acidifier (alum) to bring vase water to a pH of 3.5;
- using tepid water, about 40°C.

biocide/germicide: A chemical that kills micro-organisms that pollute water. Biocides used by flower growers and florists include
- chlorine, as sodium hypochlorite in household bleach;
- SDIC (Sodium dichloro-isocyancurate), a slow-release swimming pool chlorine;
- bromine (BCDMH) found as spa cleaning powders and solutions.

Caution Sodium hypochlorite is highly corrosive and may irritate the skin, eyes and respiratory tract. Take care to keep chlorine away from sulphur and glycerine as these may explode if in contact, even in small quantities.

Botrytis: This fungus (*Botrytis cinerea*) is also known as 'grey mould'. Its development is favoured by cool, still, humid environments and it usually attacks already-damaged petals and foliage. It is a common problem on Waxflower and Roses.

bracts: Modified leaves at the base of true flowers. They are usually green but can be brightly coloured. In Poinsettias, for example, they are the bright red 'flower', and they constitute the outer coloured parts of Proteas.

calyx: The outer whorl of flowers, composed of separate or united sepals (individual units of the calyx).

chlorosis: Yellowing of leaves or flowers, especially common in Alstroemeria and Lilies. Yellowing has been shown to relate to changes in plant hormone levels.

cleats: Rigid bars (foam or newspaper covered) put across a flower box to prevent any longitudinal movement of flowers during transport. The heads of the flowers are packed so that they are kept from touching the end of the box. To secure cleats, they are often stapled to the side of the box.

condensation: When warm, moisture-laden air hits something cool, water droplets form on the cooler surface. Therefore, after removing flowers from a coolroom it is helpful to place a plastic barrier over the flowers until they reach room temperature. This will result in condensation on the plastic instead of on the flowers and reduce fungal infection. To prevent condensation, it is also vital that the outside walls of the coolroom are well sealed; otherwise warm air passes through the outside, hits the inner wall of the coolroom (near the insulation) and then condenses, causing rusting.

conditioning: *See* rehydrating.

cultivar: Botanical name for horticultural variety or race that has originated and persisted under cultivation.

dehydration: Drying out. This usually happens to petals. Because they are so thin, it is easy for them to lose water, especially if placed in direct sunlight, draughts, near heat sources, or in air-conditioned or heated environments. Sleeving by florists is an important means of preventing dehydration. Placing sheets of plastic over buckets of flowers in the coolroom also helps prevent dehydration.

disbud: The upper-most flower that develops after all competing side buds are removed by hand. Pompom Chrysanthemums are disbuds.

dry storage: This is the best treatment for long-term cool storage of some flowers. Flowers are cooled, pretreated and dipped in a fungicide to prevent *Botrytis* attack in storage. They are then wrapped, first in paper and then plastic, to prevent drying out and stored as close to 0°C as possible. After storage, flowers are recut and rehydrated. This is most commonly done by growers with Roses prior to Valentine's Day, Chrysanthemums prior to Mothers' Day and Christmas Lilies leading up to Christmas. It is not recommended for florists.

ethylene: Naturally produced gas that causes premature ageing of many flowers, flower and leaf drop, yellowing of leaves, failure of some buds to develop, and epinasty (increased leaf angle to stem). Produced by ripening and rotting fruit, ageing and wilting flowers, cigarette smoke and car exhaust.

growth tropisms: Growth in response to environmental stimuli. The most common are
* geotropism bending away from gravity—when some spike flowers are laid horizontally they react negatively to gravity and the spike tip bends upwards e.g. Gladioli and Snapdragons.
* phototropism bending towards light e.g. Gerberas and Tulips.

MCP: 1-Methylcyclopropane. A new gas treatment used by growers to prevent the action of ethylene. Should not be used by florists.

mechanical (physical) damage: Bruising or breaking of flowers and foliage. The damage is done by grading machines (lines on Rose petals) and during transport. This is often due to flowers being handled with insufficient care when they are too open. Correct packing techniques are, therefore, very important and sleeves are essential to separate and protect bunches. Respiration and ethylene production is higher in mechanically injured plants.

microbial plugging: When flower stems are cut they exude substances that are fed on by microbes (bacteria and fungi) attracted from the water or container wall. The microbes proliferate, causing a 'soup' of bacteria and fungi to be taken up by the flower stem with water, effectively blocking the fine tubes of the water-conducting vessels (xylem). Bacteria are present on the stem and leaves of all flowers and on the rubbish/slime of vase or bucket walls. Most floral preservatives contain both germicides and acidifiers that inhibit bacterial growth. Sugar alone in water will encourage microbial growth and thus increase stem blockage.

necrosis: Death or dying of plant tissue.

panicle: Loose, branched flower-head with a large number of flowers e.g. Queen Anne's Lace.

parts per million (ppm): Number of milligrams per litre (mg/L), or millilitres per litre (mL/L), as 1 mL = 1 g. For example 100 ppm = 0.1 g/L or 0.1 mL/L).

physiological plugging: As a response to being cut, flower stems often seal over to prevent invasion by harmful microbes, similar to the way a cut forms a scab on human skin. This can effectively prevent the stem taking up water and is often overcome by acidifying the solution.

preservative (or holding) solution: Flower preservative, usually containing sugar, a germicide and an acidifier.

pulsing solution: Used on freshly harvested flowers that are in bud where a short period or 'pulse' in a high sugar solution will extend vase life or open buds.
• Recipes vary for different species, even different cultivars.
• Sugar is the main ingredient and some growers use STS to reduce ethylene sensitivity.
• Pulsing can be applied for short periods at warm temperatures, or longer periods (up to 48 hours) at low temperatures.
• Light is essential.

- A germicide must always be used if sugar is present in the solution.
- If sugar concentration is too high there may be damage, particularly to the foliage of Roses.

rehydrating (conditioning, hardening): Rehydrating is used to restore turgidity (firmness) to cut flowers.
- Rehydrating solutions contain good quality water plus a germicide and an acidifier. A wetting agent can also be added at a concentration of 0.1–0.01%.
- No sugar is used.
- Citric acid or vinegar is often used to acidify the solution to a pH of 3.2–3.5.
- A germicide is essential.
- Rehydrating can be done in the coolroom as flowers take up as much water in a coolroom as at ambient temperature.

relative humidity (%RH): The amount of water present as water vapour in the atmosphere. Most cut flowers, especially tropicals, prefer to be kept at a high relative humidity (90–95%). Humidity can be raised by wetting the floor of the coolroom, misting blooms, using plastic sleeves, or placing plastic sheets over flowers.

respiration: A metabolic process in plants that uses food (sugars) and oxygen to generate energy and heat. Usually occurs at night. If flowers are kept above about 4°C they respire faster, which creates more internal heat, further increasing respiration.

scalding: A treatment conducted by growers after harvest on flowers that contain a milky sap e.g. Poppies and Helleborus. The lower 3 to 4 cm of cut stem ends are scalded in boiling water for a few seconds. A similar treatment is searing, or burning, where the stem ends are seared over a naked flame.

SDIC: Sodium dichloroisocyanurate—granulated form of chlorine used as floral preservative and also found in slow-release swimming pool chlorine products. It is a more stabilised form of chlorine than bleach and remains active for longer than solutions made from sodium hypochlorite. Also known as DICA.

senescence: Ageing process in plants. Bud opening and the formation of colour is development, whereas flower wilting and leaf yellowing is senescence.

sleeving: Clear polyethylene plastic envelopes for flowers. Sleeving is used for:
* keeping a high relative humidity around the bunch;
* preventing tangling between bunches e.g. Gypsophila, Gloriosa Lily;
* protecting bunches from mechanical damage.

spadix: The spike of certain flowers surrounded by the spathe e.g. in Anthuriums the true flowers are contained on the centre spadix.

spathe: Broad bract or fused petal that surrounds the spadix e.g. in Anthuriums and Callas.

spike: Long unbranched flower-head with a large number of florets. Florets usually open from the bottom up e.g. Gladiolus.

spray: Composite head of flowers where all flowers are of even size after the main bud has been removed e.g. spray Carnations and Chrysanthemums.

STS: Silver thiosulphate solution. Used by growers to prevent the action of ethylene. Should not be used by florists.

sucrose: Energy (carbohydrate) source for flowers. Household sugar is sucrose.

thermometer: To monitor the temperature, keep a maximum/minimum thermometer in the centre of the coolroom away from the fan. A thermometer based at the back of your van may also yield interesting information, especially if the exhaust pipe is directly under the floor.

transpiration: The process in plants, similar to perspiration in humans. It is the loss of water from flowers and leaves to the atmosphere. Transpiration rates and water demands increase with rising temperature, with the result that the stem sometimes cannot take up water fast enough to replace what has been lost. This results in wilting.

turgid: Fully hydrated i.e. not wilted at all.

water soaking: Wet or translucent areas on petals or leaves, which often indicates chilling injury in tropical flowers; if the effect is severe then the leaves and petals collapse and dry out.

REFERENCES

The following references have been used in the preparation of this book.

Armitage, Alan A. (1992), *Specialty Cut Flowers*, Timber Press, Oregon, USA.

Dew-y's Floracare Manual, Floralife Inc., USA.

Nowak, Joanna & Rudnicki, Ryszard M. (1990), *Postharvest Handling and Storage of Cut Flowers, Florist Greens and Potted Plants*, Timber Press, Oregon, USA.

Sacalis, John N. (1989), *Fresh (Cut) Flowers for Designs*, Ohio State University, USA.

Vaughan, Mary Jane (1988), *The Complete Book of Cut Flower Care*, Timber Press, Oregon, USA.

INDEX

This book is due for return on or before the last date shown below.